$32-

From X-rays to DNA

How Engineering Drives Biology

W. David Lee, with Jeffrey Drazen, Phillip A. Sharp,
and Robert S. Langer

The MIT Press
Cambridge, Massachusetts
London, England

MIT Press books may be purchased at special quantity discounts for business or sales promotional use. For information, please email special_sales@mitpress.mit.edu or write to Special Sales Department, The MIT Press, 55 Hayward Street, Cambridge, MA 02142.

This book was set in Times LT Std by Toppan Best-set Premedia Limited, Hong Kong. Printed and bound in the United States of America.

Library of Congress Cataloging-in-Publication Data

Lee, David W., 1946–
From X-rays to DNA : how engineering drives biology / David W. Lee, with Jeffrey Drazen, Phillip A. Sharp, and Robert S. Langer.
pages cm
Includes bibliographical references and index.
ISBN 978-0-262-01977-4 (hardcover : alk. paper) 1. Biomedical engineering. 2. Medicine—Research—History. 3. Medical instruments and apparatus—Technological innovations. 4. Surgical instruments and apparatus—Technological innovations. I. Drazen, Jeffrey M., 1946– II. Sharp, Phillip A. III. Langer, Robert S. IV. Title.
R856.L383 2013
610.28—dc23
2013009442

10 9 8 7 6 5 4 3 2 1

This book is dedicated to my first wife, S. Ramsdell (Ramsey) Lee, whose fight against cancer opened my eyes and inspired me to become involved in cancer research and medical device development.

Contents

Preface

Born in the 1940s, I was one of those inexhaustibly curious children who built a laboratory in the basement of his house in Detroit. Complete with natural gas Bunsen burners, stocked with nitric and sulfuric acid and magnesium ribbon, I "experimented" with real chemistry. One very cold winter day, after filling the house with billows of chlorine gas (which forced the whole family to evacuate), my parents decided it was time for me to graduate to electronics and motors. My biggest accomplishment was building a motor that powered nothing!

My first exposure to organized scientific development was in 1966, when I spent the first of several summers as an "accelerator technician" at the Lawrence Berkeley Radiation Laboratory, known then as the "Rad Lab." It was invention and building experiments on a super grand scale and I was in my element there, surrounded by cyclotrons and other magnificent gadgets. In 1969, after graduating from MIT, I was hired by Arthur D. Little (ADL), a large, prestigious, technology-based consulting firm, where I was ultimately in charge of the technology and product development business. During my 25-year tenure there, we built from scratch nearly everything we worked on (cryogenics, combustion, appliances, space hardware, etc.). A principal investigator at ADL (called a case leader) would begin a project with a lab containing only empty benches. He or she would then proceed to design and build the equipment that would be the basis of the team's research. At ADL, I learned how to conduct research from Dr. Robert Wilson, Dr. Joan Berkowitz, and Dr. Peter Glazer. Bob taught me how to think through the design of an experiment, constantly refining the understanding until it could be summarized on a single page or, better yet, a 3 × 5 card. Joan taught me the ins and outs of chemistry, which I seemed to have missed in my years at MIT. Peter Glazer showed me how to reach out beyond the obvious and conventional and have the courage to propose new technologies; then relentlessly pursue funding until you are able to build what you dream.

After losing my first wife to cancer in 2003, I made a career change from hardware development to cancer research. I had the good fortune to know Professor Tyler Jacks, the director of the MIT Center for Cancer Research (CCR). Taking a huge risk, Tyler invited me to meet others at CCR. I had further luck in that MIT was in the process of formulating the concept of integrating engineering and biology in an effort to address cancer at what is now the MIT Koch Institute for Integrative Cancer Research. In 2008, I was offered a role at CCR to help with the interface between biologists, engineers, and clinicians. As I was being introduced around CCR, I was struck by the uniformity of the different biology labs. Of course, at that point, I did not understand enough of the subtleties to see the differences. I could not help comparing my then simplified view of biological research with my exposure to research in high-energy physics and so forth, where the key research tools were always purposefully built by the team concurrent with the design and conduct of the research. This seemed like a dramatic difference, and I questioned its importance. I looked for case studies that explored the integration of engineering and biology in an effort to accelerate the discovery and found nothing. I chatted with my friend Dr. Jeffrey Drazen, the editor in chief of the *New England Journal of Medicine*, about his observations, and he encouraged me to do my own primary research. Simultaneously, Phillip A. Sharp was writing about the convergence of the life sciences, physical sciences, and engineering. Conversations with him helped shape the theme of the book, and I was off and running.

Acknowledgments

It has taken many years, many minds, and much encouragement to get here.

From the very beginning of this project, I have had the unwavering support of Dr. Jeffrey Drazen (editor in chief of the *New England Journal of Medicine*) and two MIT Institute professors: Dr. Phillip A. Sharp and Dr. Robert S. Langer. The three brought such unique perspectives that I decided to include interviews with them in the book. I can say without hesitation that without Jeff and Phil's encouragement and time-consuming involvement over the past 3 years, I would not have completed the effort.

Unlike typical acknowledgments of wives who suffer the months and, in this case, years of neglect while the writer writes, my wife, Eve Youngerman, had the foresight and wisdom to choose to become an integral part of the effort and write a large part of one chapter. Eve was a consistent and reliable cheerleader. She was assisted in her research by our daughter, Ruthie Lewis, an undergraduate studying biochemistry.

After the initial draft, I worked with James Buchanan (Orchard Writing) on the structure, flow, and content of the book. James made big contributions throughout the book helping to craft more readable pages and improving the flow of the ideas. James was also a good teammate for when it seemed as though the book would never be completed: He was a positive force.

Along the way I had the fortune of receiving insight regarding themes and approaches from Dr. Andrey Zarur, managing partner of Kodiak Ventures, as well as from Dr. Robert Urban, then the executive director of the MIT Koch Institute and now at Johnson & Johnson. Andrey could in a matter of minutes picture the entire subject and suggest important thematic changes. I always value Andrey's suggestions. At the first complete draft, I enlisted my brother Kevin Lee, an attorney, to read the entire document. I have leaned on Kevin my entire life starting back on the playgrounds in Detroit so it seemed like a logical thing to do. I wanted to see if someone outside of the medical community would be interested in the material. He was, and he then made important changes to the historical analysis. At the same time I got editorial comments from my longtime friend Terry Finn, also an attorney and outside of the medical community.

Once the draft was completed, I had the benefit of expertise from some talented people. Dr. Annetine Gelijns is the professor of health policy at the Icahn School of Medicine at Mount Sinai and a researcher and scholar on the subject of technologies in medicine. Annetine helped me correct some important logical errors and clarify my observations. I only regret not having started our collaboration earlier. Dr. Nadya Dimitrova, a postdoctoral student in the Jacks Labs at MIT, poured through the entire book, ensuring that the biology discussions were correct and consistent. Nadya did a remarkably thorough edit, which surprised me by the sheer number of corrections. Robert Lewis, my son and a Ph.D. candidate, also made suggestions and edits to improve the book.

While my two daughters, Katie and Merritt, did not research or write anything, their beaming smiles when I talked about my research findings for the book were a constant inspiration. My other son, Jamie, and my two sons-in-law, Jeff Fishbone and Kevin Garofalo, listened patiently as I droned on about my epiphanies. The biggest thanks, however, must go to my parents, Barney and Mary Lee. Their support of my insatiable childhood curiosity, encouragement of my early scientific explorations, and unconditional belief in my abilities have allowed me to achieve the successes in my life.

I

INTRODUCTION

1 An Opportunity for Greater Discovery

For many in the developed world, the word *tuberculosis* conjures an image of a long-vanquished infectious disease akin perhaps to polio or smallpox. Tuberculosis is a disease that is beyond the memory of most, but it is a term loaded with potent symbols of its past virulence: sickened patients coughing up bloody sputum in sanatoriums.

The suffering caused by tuberculosis has vanished in the modern age, as have the many sanatoriums (isolated facilities for tuberculosis patients), which have been converted to historical landmarks or dedicated to other uses. To people in the United States and other developed nations, the fear produced by the threat of tuberculosis epidemics is a piece of historical ephemera.

By contrast, in developing nations such as Haiti, tuberculosis is a current and all too real fear. Prior to the earthquake in 2010, tuberculosis was second only to HIV/AIDS in terms of the number of deaths caused by infectious disease. In 2007, nearly 7,000 people in Haiti died of tuberculosis, and more than 29,000 Haitians developed the disease in 2009. Most of the country's infrastructure, which included numerous medical facilities, was destroyed. This in turn led to a diaspora of patients seeking shelter in numerous densely packed camps in and around the country's few cities. Countless others also left the ravaged population centers to find succor among their hometowns and villages. Making matters worse, without consistent access to treatment, sufferers likely developed and passed on drug-resistant strains of the disease.

In many ways, the earthquake took Haiti back to a time before modern health care systems had the tools properly to diagnose, treat, and cure this disease.

At the beginning of the nineteenth century, tuberculosis, known then as *pulmonary consumption*, was an endemic infectious disease. It could only be identified in its later stages after its most severe symptoms—bloody cough, night sweats, and extreme fatigue—clearly demonstrated its presence as distinguished from a number of other communicable diseases.

Medicine of the time simply lacked the tools to diagnose tuberculosis and the medicines to treat it effectively. This meant that each day was the equivalent of the tuberculosis diaspora of Haiti. Imagine living at a time when a simple ride in a horse-drawn carriage across town or to another city could bring you into contact with an individual whose only symptom is a persistent cough. As annoying as it may be, it would not be inconsistent to encounter many people each day with some sort of chronic or acute pulmonary ailment.

After arriving at your destination, you wish your fellow travelers well and then walk to work or home to your family and quickly forget the incidents of your ride and your companions. The only memory of it is captured in the tuberculosis bacteria you inhaled after your fellow traveler coughed and did not properly cover his or her mouth.

Over the next few weeks, months, or perhaps longer, the bacteria take up residence in the tissues of your lungs. In this fertile environment, they grow and multiply causing at first minor symptoms. Before long, though, you have developed a persistent cough that begins to produce bloody sputum. You have unaccountably lost weight, experience night sweats, and are bone tired.

You consult a physician, who, based solely on the report of these symptoms, diagnoses you with pulmonary consumption. The stethoscope has not yet been invented, so he does not have the ability to perform even a cursory examination of your lungs and heart. The notion that it is caused by bacteria is completely unknown to him. If you mention the coughing companion of your carriage ride, it would hold little meaning for your physician because he knows little about how the disease is contracted.

What he can tell you is that there are no effective treatments. If you have the money, you can seek comfort at a sanatorium, but most likely you return home to your family. Although some diagnosed with consumption manage to survive, it is very likely you will soon die. It is also highly likely that your family and friends, whom you have most certainly infected, will suffer the same fate.

Half a century later, the fate of the average consumption patient began to brighten. Working with one of the first compound microscopes sensitive enough to detect bacteria, Robert Koch identified the bacterium that causes tuberculosis. About 15 years later, the development of the Crookes tube—the device that helped identify electrons—led Wilhelm Roentgen to discover X-rays.

A few years later, X-ray technology was made commercially available to doctors and hospitals, which allowed doctors to diagnose tuberculosis before it became florid. This in turn allowed organized medical systems to better control outbreaks of tuberculosis. The disease was still deadly and pervasive, but medical science had begun to gain traction against the disease.

By the 1950s, new tools had been developed to diagnose the disease better. Other tools allowed researchers to culture the bacteria and to develop effective treatments to cure it. With effective diagnostic tools and treatments in hand, large-scale public health measures were undertaken to prevent infection and to diagnose and cure it when it occurred. As a result, the disease burden was reduced to the point where the continued existence of sanatoriums was threatened by a lack of need.

Even still, diagnosis could take weeks to months to confirm, during which time the patient could be infectious to others. Treatment was effective, but difficult.

Now, move ahead in time to 2012. Using modern molecular techniques, we not only can make the diagnosis in a matter of hours but also can tell if the isolated bacterium carries genes associated with a drug-resistant signature of tuberculosis. Additionally, treatment has been greatly improved, though it can still take 6 to 9 months to cure a patient fully.

As the examples of Haiti and other developing nations have shown, in 200 years tuberculosis has gone from a problem of cure to one of delivery. At each step of this journey, technologies such as the stethoscope, microscope, X-ray, molecular techniques, and many others have played a critical supportive role to advance biological research.

Therein lies the premise of this book. Engineering has been an essential collaborator with biological research leading to significant breakthroughs throughout all fields of medical science. Often, the engineering breakthrough is achieved in a setting that is removed from the biologist. The two are only brought into collaboration after the technology has been brought to market. The result is that a time period of 20, 30, or 40 years will elapse between when the technology is developed and when it is available for the biological discovery. This delay is a key metric we have quantified in this book.

Major advancements of our understanding of biology (e.g., the structure and nature of DNA) can be understood through Thomas Kuhn's book *The Structure of Scientific Revolutions*, where he shows advancement as leaps, or paradigm shifts. He traces scientific discoveries to technologies that produced a result that was an anomaly [1] within the current paradigm and for which the scientist had the insight to recognize the data as a true inconsistency. It is the new technology that produces the new data that produces the new insight and the advancement.

The author suggests that enabling technologies are a pacing item in the advancement of our understanding of biology. Accelerating the engineering and availability of the technology to the biologist will move the science more rapidly, leading to swifter enhancements in public health as well as professional achievement among researchers and research institutions.

In many ways, the most apt modern corollary to the battle against tuberculosis is cancer. By tightening the bond between engineering and biological research, it seems likely that we can reduce the gap between technological development and medical breakthrough from 40 to 7 years if not less. Hopefully, the time it takes to diminish cancer from a problem to a cure will be greatly reduced.

Engineering: Critical Research and Clinical Partner

To tell this story, the author will identify key inflection points in the development of the life sciences and then analyze the critical factors that have converged to make breakthroughs possible. However, the focus is not purely or even primarily historical.

By exploring and interpreting past innovations, the author will suggest a methodology for addressing some current pressing needs in health care. This book will examine cancer as an example to demonstrate how future breakthroughs might be nurtured and accelerated through a tighter, more collaborative coupling of engineering with biological research.

In essence, the author is seeking to present a thoughtful argument for a fundamental change in how this field of research is structured and funded. The historical record shows that important leaps forward in the life sciences require a combination of innovative scientific/biological work and one or more critical enabling technologies. These technologies, or tools, are the product of skilled engineers.

For example, decoding the double helix structure of DNA—an elegant biological breakthrough—only became possible when the engineering community made significant advancements in several technologies, such as X-ray diffraction and gel electrophoresis. As but one example, this major leap forward in our understanding of the building blocks of all life was made possible and propelled by the introduction and commercial availability of key technologies. The biologist (or biophysicist) use of X-ray diffraction had to await the engineering transformation of X-ray technology into a commercial instrument over a 40-year period..

The importance of bringing engineering disciplines into the core of biological research and clinical practice is not a novel concept. It has been underscored in many quarters. Philip A. Sharp—winner of the 1993 Nobel Prize in

Physiology or Medicine—has called the infusion of engineering into biological research a key revolution in biology.

The promise of this revolution is the rapid translation of new molecular and cellular knowledge into diagnosis and treatment through engineering approaches. There is ample opportunity to achieve greater advancement of cellular research through new nanoscale quantitative methods and measurements developed by engineers. Parallel to these developments will be advancements in drug development and engineered nanotechnology for precise drug delivery.

This rate of advancement within the realm of biological research is entirely achievable via *concurrent engineering*, the author's preferred term for the convergence of engineering and biological research. However, adaptation of current structures to this methodology is inherently difficult and will require us to develop new funding and perhaps teaming mechanisms.

Why? Biology and engineering are fundamentally different cultures. As interviews later in the book tell us, they are funded differently, they approach the exploration of basic science and the formulation of problem solving differently, and, most importantly, engineers and biologists think differently. Most biology researchers—and those that fund them—do not encompass the development of new enabling technologies in their work. Most limit their research to use of technologies that are already commercially available, which acts to constrain their field of inquiry.

One has to wonder how many times a scientist with a compelling hypothesis has the nagging thought, "If only I had the right technology."

So, how can this new kind of collaboration be made more common and more successful? What can we learn from past biological discovery and from related advancements in the clinic? How can funding organizations more effectively stimulate breakthroughs in our understanding of biology, disease, cures, and wellness?

Drawing on real-life case studies, this book will reveal a stepwise progression of technology development enabling medical advancement. Finally, and most importantly, the author will suggest how to accelerate life sciences advances that are critically required to support new health care and wellness paradigms.

We focus most of the book on breakthroughs or new paradigms in biology and devote a couple of chapters on engineered medical technology in the clinic. What we do not address are the complex issues of bench-to-bedside transition. As Dr. Jeffrey Drazen (contributor and advisor) put it: "You make scientific progress, but making medical progress is another order harder. It is like we understand, I think, at the mouse level a lot of basic biology and how do I then transform that to a therapy or to a diagnostic? Because we now have to deal with all the variance that comes from people being different people, both genetically and in their environmental exposures. With this biological variance our simple, and often elegant, understanding is confounded. What we have learned from the mouse does not accurately predict what will happen in people, and our attempts at therapeutics or diagnostics fail to meet the needed clinical tests."

2 Concurrent Engineering and Science

To set the stage for our discussion of concurrent engineering, it is helpful first to look at an area of scientific endeavor where the marriage of engineering and science has had a profound impact. Although there are many such examples to choose from, the personal experience of the author leads us to examine the unraveling of the structure of the atom and research on the planets, stars, and moon.

As a young man, the author was fortunate enough to have been an engineer supporting leading-edge scientific discovery at the Lawrence Radiation Laboratory in Berkeley, California. This experience as an integral team member for a range of scientific explorations laid the groundwork for his understanding of the role engineering plays in the advancement of scientific research.

Initially, his work centered on developing measurement technology for high-energy particle experiments while at Berkeley. After graduating from MIT and joining the staff of the engineering consulting firm Arthur D. Little, he transitioned to the role of test engineer for several lunar experiments that were carried to the moon on Apollo 15 and Apollo 17. This work included development of the Traverse Gravimeter and the Lunar Heat Flow Probe.

As a side note, another mission-to-the-moon technology called the Laser Ranging Retro Reflector unit (which the author was not involved in though it was designed and built at Arthur D. Little) is the only remaining experiment on the moon that is still operational to this day. It is also important to note that during his research for this book, the author could not find any public reference identifying the engineering team that designed, built, and tested this remarkable technology that is still functioning on the moon. Unless you knew them already by name and dug into their relatively obscure personal histories in archives, you could not identify the people who actually made it all work. You can easily find the names of scientists whose research was enabled by the technology.

While the author's interest in the role of the engineer in scientific exploration may seem somewhat tangential, it in fact holds relevance to the core argument of this book. The anonymity of the engineer who conceived, designed, and built the technology that enabled the scientific breakthrough is an issue found throughout our historical review. This will be discussed in greater depth later, but it is important to note that the anonymity of engineering leads the public and research funding sources to undervalue the important role it plays as a concurrent partner to scientific discovery.

The Dawn of Physics: From Mystery to Determinism

The nature of the atom and its behavior was largely unknown prior to the gigantic discoveries of the late 1880s that revealed the structure of matter. Before that time, the fundamentals of the structure of matter were a mystery. The behavior of light and matter were observed but unexplained.

Additionally, there were physicists who believed the physical world was not deterministic. They disagreed with the concept of natural laws that could not only be discovered and described by science but also used to understand and predict the behavior of light and matter.

The means to resolve this disagreement was not through generation of more data via the classical framework with the classical tools. Rather, it was the development of new technologies that would reveal the subatomic and quantum world and advance the existing framework of understanding.

The late 1800s mark a turning point because brilliant physicists began to create and/or use emergent technologies to devise ingenious experiments. These tools and experiments revealed the components of the atom and subatomic behavior, and the mysteries of the atom began to disappear.

Engineering-Enabled Science: Quantum Physics, 1880–1930

During this period of discovery, brilliant and creative physicists such as Ernest Rutherford, Robert Andrews Millikan, Albert Abraham Michelson, Heinrich Hertz, Arthur Compton, and others pushed the envelope of our understanding of the quantum nature of atoms. The keys to their success were magnificently conceived experiments that they and their collaborators engineered and built.

Rutherford, Millikan, Michelson, and others performed both engineering and research physicist roles; there was essentially no time delay between the engineering of the experimental technology and its application to research and discovery. The following case studies will show that physics leapt ahead because engineers were working concurrent to the scientific endeavor and the physicists themselves were comfortable tackling engineering challenges.

The Start: X-Rays

Around 1870, William Crookes engineered an evacuated glass tube with electrodes at either end. The anode was perforated to allow electrons to pass through it and excite a fluorescent dye painted on one end of the tube behind the anode. The early significance of the development of this technology was the discovery of cathode rays—also known as electron beams—which are streams of electrons.

The Crookes tube, as it became known, also supported numerous experiments as it was relatively inexpensive to make and performed reliably.

Some 30 years after the development of the Crookes tube, now a mature technology, Wilhelm Roentgen was experimenting when he noticed an as yet unobserved phenomenon. One day, he was operating the Crookes tube with a piece of black-painted cardboard set between the end of the tube where the electron stream originated and the surface painted with fluorescent dye. He had expected that the cardboard would block the electron stream thereby preventing the electrons from exciting the fluorescent dye and making it glow.

Instead, the fluorescent dye glowed faintly. Surprised by this discovery, he began to explore the properties of the Crookes tube with more experiments. Ultimately, he concluded that the Crookes tube was generating some type of ray that could travel through the cardboard and excite the fluorescent dye.

He then persuaded his wife to place her hand between the mysterious ray and a surface coated with fluorescent dye. The result was a photograph-like image showing the bones of his wife's hand along with the ring on one of her fingers. The X-rays, as he called them, passed through the tissue but were blocked by the bones and the metal of the ring to create a fairly detailed image of the hand.

Roentgen's discovery of X-rays was enabled by the earlier development of the Crookes tube as well as by its wide availability to the community of scientists exploring the evolving world of physics. X-ray technology, of course, remains to this day a critical tool in the arsenal of medical equipment used for the diagnosis and treatment of people.

Discovering the Nucleus of the Atom

Following on the heels of the X-ray discovery, Ernest Rutherford and his two co-workers, Hans Geiger and Ernest Marsden [2], were immersed in examining the behavior of alpha particles on different materials. Alpha particles are one of three types of radiation—beta and gamma particles being the other two—and are considered to be a weak form of radiation.

Rutherford and his team hypothesized that if they could observe the scattering of alpha particles after passing through a very thin sheet of gold, they could ascertain the structure of the atoms contained in the gold. The idea was similar to shining a light on your hand and projecting it onto a wall. The shadow would outline the general shape of the hand. If your hand were of atomic scale, the larger shadow would enable you to see it is composed of five fingers and a palm.

To detect the scattered alpha particles, they built an elegant apparatus with a zinc sulfide screen placed behind the gold foil.

Quite unexpectedly, they found that most of the alpha particles passed straight through the gold foil and impacted the zinc sulfide screen in a small area. However, every once in a while some particles would be detected to the side or even reflected back toward the source. This was unexpected and could only be explained by a structure in which most of the atom weight was concentrated in a small nucleus with surrounding electrons.

They had discovered that atoms are composed of a nucleus and surrounded by an indeterminate number of electrons.

As an example of concurrent engineering, Rutherford, Marsden, and Geiger designed and developed their own experimental apparatus to then conduct their research. To these scientists, having the engineering skill to conceive and develop the research instruments to explore their area of interest was just a natural part of their research effort. The time from development of the engineered technology to the time of discovery was nearly zero.

Measuring the Charge of Electrons

Continuing the exploration of the atom, Robert Millikan and Harvey Fletcher measured the elementary electric charge of an electron in 1909.

Their concept was to create an apparatus that could suspend charged oil droplets between two charged plates. The electrostatic charge of the oil drop caused it to be attracted to the upper plate when a voltage was applied, and when the voltage was removed it would drop.

Using an X-ray in the chamber, which was known to charge the oil droplets, and then adjusting the voltage sent to each plate, Millikan and Fletcher could slow the fall of the droplet between the plates. By running the experiment on numerous droplets, it was expected they would see a slightly different number of electrons due to droplet size variation. They could then calculate the difference between the droplets.

As the experiment unfolded, it became apparent that the charge differences were all multiples of a small fixed calculated value. This, they concluded, was the discrete charge of an electron.

Table 2.1
Technologies and associated discoveries

Enabling Technology	Nobel Prize–Winning Scientific Discovery
Crookes tube, 1869	Roentgen: properties of X-ray, 1895
Rutherford, Marsden, and Geiger alpha-scattering apparatus, 1908	Rutherford: structure of nucleus, 1908
Millikan and Fletcher oil drop, 1909	Millikan: electron
Michelson interferometer	Michelson: speed of light
Karl M. G. Siegbahn X-ray spectrometer	Siegbahn: spectral lines of elements
James Franck and Gustav Hertz apparatus, 1914 [4]	Franck and Hertz: confirmation of Bohr quantum theory
Clinton Davisson and Lester Germer electron diffraction, 1927	Confirmation of de Broglie wave theory of matter

Like Rutherford, Millikan was skilled in atomic physics and mathematics as well as engineering principles, so that he was able to design a tool that repeatedly produced meaningful measurements. For Millikan, the time difference between the engineered apparatus and its use in the fundamental scientific discovery was essentially zero.

The above discoveries—all of which led to the awarding of Nobel Prizes—were enabled by the creation and availability of engineered technology. With the exception of the Crookes tube, each of these physicists engineered his own tools, which led to a time gap between engineering and discovery of zero.

The list in table 2.1 summarizes the enabling technology that was developed concurrently with the Nobel Prize–winning scientific discovery. This is a powerful testament to the importance of concurrent engineering not just to discovery but also to the ability of these men to make their mark in their profession. The history of the physics prize was taken largely from the Nobel Prize Web site [3].

The physicists at the beginning of the twentieth century had an engineering hat on when building the enabling technology and a physicist hat on when designing the experiments and developing the governing relations. In fact, our survey of Nobel Prizes in Physics awarded between 1900 and 1970 showed that about 80% involved the engineering of an experimental device. Of these, 40% were engineered by the physicist. The convergence of science and engineering was achieved by each of these men in their own work.

These physicists moved the frontiers of science by designing and building technology that uniquely allowed them to discover the phenomenon they were after. And they did this without any formal training in engineering.

Physics at Lawrence Radiation Laboratory, 1930 to Today

The nature of physics and engineering convergence began to change around 1930. To explore the atom beyond its basic structure, scientists required tools capable of penetrating the atom and exposing its components. This meant generating higher-energy particles that could smash into the nucleus to separate it into its constituent pieces.

A cyclotron is the technology that was developed to accelerate charged particles so that they travel in a spiral path out from the center until they smash into a target of interest. Each "dee" holds a magnetic field, and the gap between them contains a rapidly varying electric field. The transference of the particles around the dees and across the electric field creates a significant amount of force. As the particles cycle through the gap from one magnetic field to another, they accelerate.

Ernest Orlando Lawrence, winner of the 1939 Nobel Prize in Physics, built the first cyclotron instrument used to explore the nucleus of the atom. Figure 2.1

Figure 2.1
Ernest O. Lawrence examining a 37.5-inch vacuum chamber with the lid removed, circa 1935. Photo courtesy of Berkeley Lab. © 2010 The Regents of the University of California, Lawrence Berkeley National Laboratory.

shows Lawrence with the 37.5-inch cyclotron in 1935 [5]. Shown are the two metal dees (the two hemispheres of the cyclotron) and the narrow gap between them. The cover is off the cyclotron.

Another magnet located on the periphery of the cyclotron pulls the accelerating particle out from its central orbit thereby creating the outward spiral path. The particles eventually spiral out an aperture as a high-energy beam and then collide with a target of interest.

To push the scientific frontier, physicists needed to impart higher energies to the particles under study or to accelerate larger particles, so the size of the machines grew as well. Engineers like William Brobeck [6] joined with applied physicist Lawrence and theoretical physicist J. Robert Oppenheimer, and together they built the first large-scale cyclotron (figure 2.2) [7].

These big machines gave way to still larger particle accelerators (cyclotrons) such as the billion-electron-volt bevatron and linear accelerators. Figure 2.3

Figure 2.2
The 184-inch cyclotron converted from a calutron to a synchrocyclotron in 1945–1946. Ernest O. Lawrence and staff posed with the magnet. Photo taken April 28, 1946.
Photo courtesy of Berkeley Lab. © 2012 The Regents of the University of California, Lawrence Berkeley National Laboratory.

Figure 2.3
A rare view of the HILAC. This photograph, taken near the completion of the modification program, shows details usually hidden by shielding. Poststripper tank is in the foreground. Photo courtesy of Lawrence Berkeley National Laboratory. © 2010 The Regents of the University of California, Lawrence Berkeley National Laboratory.

shows the Heavy Ion Linear Accelerator (HILAC) in 1965 [8]. The HILAC was the apparatus that synthesized entirely new synthetic elements known as transuranic elements. These are highly radioactive elements with atomic numbers greater than 92 and include neptunium, plutonium, americium, curium, and californium and resulted in Nobel Prizes for Edwin M. McMillan and Glenn T. Seaborg.

Parenthetically, around this time (1940) at the same facility—the University of California at Berkeley Radiation Laboratory—Martin Kamen and Sam Ruben discovered carbon 14. This is the radioactive isotope of carbon that ushered in the further discovery of many radioactive isotopes that became the backbone of tracers used in countless biological discoveries. These will be discussed in subsequent chapters.

The integration of science and engineering to explore the atom is self-evident in the progression of the first accelerator to the ones that now sit atop the Berkeley campus housed in their own buildings. Additionally, the work in high-energy physics that spawned the development of cyclotrons and linear accelerators stretching more than a mile created a scale of science that became known as *Big Science*. We will expand the discussion on this subject when we consider Big Science in biology in chapter 12. Importantly, these Big Science technologies allowed scientists to explore the structure of matter to a level that was impossible before the availability of these technologies.

For now, though, one can point to a number of major discoveries at Berkeley that were enabled by engineering. In the 1960s, the newly named Lawrence Radiation Laboratory hosted the discovery of the elements lawrencium (103), dubnium (105), and seaborgium (106). The unearthing of these elements at Berkeley was largely dependent on the availability of the cyclotron sitting atop the hilltop overlooking the Berkeley campus and the San Francisco Bay.

The cyclotron was an engineering marvel in its own right and continues to operate and provide important scientific information. However, to conduct the science requires more than just the accelerator. Sophisticated devices are needed to create and measure the interactions that reveal the composition of the atom.

For example, a high-energy beam is extracted from the cyclotron by very large and powerful quadrupole magnets that impel the accelerating particles into an outward spiral and then into the target. The high-energy particles that smash into the target and the resultant tracks of the particles are analyzed to reveal the atomic composition of the original target.

In the early days, the physicists and engineers at Lawrence created huge liquid hydrogen bubble chambers cooled to –423°F. The bubble chamber consisted of a bathtub-sized structure filled halfway with liquid hydrogen. The hydrogen was maintained at a temperature and pressure just at/below its boiling point and would show the trails of the particles resulting from the collision of the high-energy beam with the target of interest as they passed through the cloud of dense vapor created by the hydrogen.

Maintaining the hydrogen vapor at –423°F required a complex assembly of cryogenic plumbing and instrumentation. The entire bubble chamber structure with its plumbing and redundant safety systems was the size of a typical house and had to be placed accurately behind the target of interest. Physicists and engineers with both structural and cryogenic expertise worked together as a team to create the first hydrogen bubble chamber, which then opened the science of energy physics at the Lawrence Radiation Laboratory.

While the bubble chamber was a sensitive and effective detector of subatomic particles, the analysis of the data was cumbersome. High-speed film was used to track the particle trajectories in the hydrogen vapor, and then a room full of people called "scanners" looked at the film. As the cameras produced miles of film with each experiment, the analysis generally took more than a year.

Another approach was needed. In the mid-1960s, a new team of engineers and physicists was assembled to find an innovative and far more efficient way forward.

The concept they developed was based on allowing the highly energized particles to enter a box with a perforated set of charged plates that resembled something akin to a wire screen. The wires were charged and would experience a small electric impulse or spark as the particles ionized the gas as they passed through.

Magneto collective strips around the periphery and attached to the wires would detect the specific wire that experienced the ionization pulse. This provided an instantaneous X/Y position at a precise moment in time. The next plate of wires behind the first would detect the pulse's location in the next instant in time.

This wire chamber—or spark chamber as it was known—gave instant test results, which eliminated the time-consuming and error-prone method of scanning the bubble chamber films.

Both the bubble and spark chambers are excellent examples of the power of concurrent engineering. The development and planning of these chambers were engineering projects accomplished by a team of physicists and engineers addressing the scientific challenge of measuring, monitoring, and observing high-energy particles.

Engineering and Astrophysics: Creating the Hubble Space Telescope

One of the best-known collaborations of engineering and science is the story of the Hubble Space Telescope and the team that transformed it from a hopeful vision to a powerful reality. At the heart of this story is Dr. Rodger Doxsey, a close personal friend of the author who passed away in October 2009 at age 62.

To understand the connection of the Hubble Space Telescope to this story about engineering supporting scientific discovery, the author interviewed Dr. Matt Mountain, the director of the Space Telescope Institute, on April 30, 2012. Here, he starts the story with the early proposals for a space telescope:

> The most famous is the 1947 paper by Lyman Spitzer when he was working for the Rand Corporation, and the space program, as you'll note, hadn't really started, and he realized from a sort of theoretical perspective, if you put telescopes in space it would open up the whole field, but he had no idea how to do it or make it happen. Now move time forward, and the DOD realized, hmm, if you put the telescopes in space and look

down, you know, you can do these things, and the Hubble was probably the fourth or fifth in that series, so astronomers finally got their first telescope in 1990.

I think astronomy traditionally, observational astronomy, has always been enabled by technology development, and on the flip side, if you're lucky, the other side of that, it's also had to push technology to make it happen. . . . there were two traditions in astronomy. There is the physics side, which is where Rodger came from, and what I would call the traditional astronomic side, which is more the observer.

And those two traditions have been intertwined, you know, over 400 years, but I think it's really on the physics side where we've seen most of the intertwining of engineering and astronomy.

[T]he thought that you would just buy technology off the shelf and then you wait for the technology development, it's just so alien to our culture that I just—you know, it's not a part of the way we think about these things. When I had the technology, you know, off the shelf, it isn't particularly useful.

Well, I think it's very important—I just want to make a clear distinction. The X-ray team was basically a bunch of physicists who I think would invent new technology to get their science done, right? Infrared astronomy, radio astronomers were the same. The more traditional optical astronomers from the fifties on had telescopes provided to them and were much more traditional in that view that they were observers. They took the telescopes they were given and a certain conservatism set into that community.

It is in this setting of an intensely scientific investigation of the origins of the universe with a new enabling technology of the space telescope that a master of both disciplines, Dr. Rodger Doxsey, emerges. Rodger understood the science having received his B.S. and Ph.D. degrees in astrophysics from MIT. He also understood the hardware and was master of the complex computer program that actually runs the telescope on a minute-to-minute basis. Parenthetically, he and the author rowed crew together as undergraduates and thereafter. Dr. Tod Lauer, an astronomer on staff at the National Optical Astronomy Observatory in Tucson, Arizona, said that Dr. Doxsey played a critical role in the early days of the telescope. "In the early days of the telescope it was a rather crabby and cranky spacecraft and Rodger was very adept at getting it to work properly," said Dr. Lauer. "In simple terms, he figured out how to get the trains to run on time during a period when it wasn't clear if the train was even on the track" [9].

Later, Dr. Doxsey's expertise was called on once again when NASA administrator Sean O'Keefe made the decision in 2004 to end the astronaut servicing missions. This policy was instituted as NASA began phasing out the space shuttle program and left Hubble's engineers and astronomers scrambling for ways to prolong the telescope's life. Among the issues they had to find solutions for was that the gyroscopes, which help the telescope find and keep itself fixed on points in space, had a habit of failing.

Working closely with engineers at the Goddard Space Flight Center in Green Belt, Maryland, Dr. Doxsey and his team developed a way to save the fragile gyroscopes. Up to that point the telescope required all three of its gyroscopes to function properly in order for it to find and fix on a point in deep space. This is an intensely complicated and precise piece of engineering and science on its own.

Instead, Dr. Doxsey and his team devised a method of operating the telescope with just two of its gyroscopes and then filling in the missing information from other spacecraft systems. This ingenious and complex solution enabled the Hubble Space Telescope to perform as well on two gyroscopes as it had on three. In May 2009, new components were sent to the telescope aboard the space shuttle *Atlantis*. As these pieces of equipment were installed, repairs were also made that have extended the life of the telescope even further.

Once again, Matt Mountain picks up the storyline. "I mean Rodger was one of the smartest people I ever knew, and he came from the implementation X-ray background, so he came from a very similar background that I did. Rodger had a very deep technical understanding of the telescope, more than anybody else, as well as the science. That was a very unique skill for him. He really did straddle that world. And those kinds of people make progress because without the technology, you don't make progress."

There is no doubt that the combination of engineering with the disciplines of astronomy and astrophysics produced a tool of incredible power. The engineering made the tool and technology possible, while the scientists were enabled to make discoveries of everlasting importance and grandeur.

In no small terms, this is the promise of concurrent engineering as it could be applied to biological research.

3 Engineering and the Engineer

Early in the writing of this book, the author asked Dr. Robert S. Langer, the David H. Koch Institute Professor at MIT's Department of Chemical Engineering, to review part of this book. After reading a handful of the stories that follow this chapter, Dr. Langer asked, "Where are the engineers?"

In part, this is the subject of this book: placing engineering within its proper role in the context of biological discoveries. To date, while the handprint of engineering can be found in most of the biological discoveries, the identity of the engineers are largely undiscoverable—they remain the nameless sources of the enabling technologies.

After some months of work on the subject of the missing engineer and the role of engineering in biological discoveries, two conclusions became obvious.

First, a large number of biological and clinical advances were enabled by technology only after the technology had been engineered and offered as a commercial product. Generally, recognition flows to the researcher who uses the technology and not to the creative technologist who made the technology possible.

Second, the creation of an enabling technology generally occurs in two major steps. The first step is a proof of concept (POC), where the scientific principles (the author calls them first principles) are discovered and demonstrated either in hardware or mathematically. The POC stage is most often driven by a scientist engaged in chemistry or physics research. For example, Isidor Rabi (see chapter 13) built a device that was designed to confirm the nuclear magnetic resonance (NMR) of a specific gas but could not be used to characterize biological materials and certainly not liquids or solids. The second stage is the product development phase (PDP), which itself might be broken into alpha, beta, and product prototype stages. Generally, this stage does not involve new science of new first principles but is often a very creative process of combining physical principles into a working instrument that the broader research community can use. Twenty years after Rabi's effort, Russell Varian and others at Stanford translated Rabi's findings into a device that could be operated as a useful instrument on a variety of materials. They went on to commercialize the NMR spectrometer.

During the research and writing of this book, the author found that in nearly all cases, the POC is done in academia with resulting publications and recognition of the principle investigators. The PDP, in contrast, by its very nature is done in industry and generally takes place without publication or recognition of the individuals (Varian was one of few exceptions).

Further, by digging deeper into the published material on the development of the first principles and the POC, the author hit upon another insight. Even though the technology development work was generally done by individuals with degrees in physics or chemistry, the development work itself was engineering.

Engineering Defined

To say the least, engineering is a diverse and vast area of endeavor that works on a scale of nanotechnology to the largest human-made structures on the planet and beyond. However, its principles are most often defined as: *The application of scientific and mathematical principles to practical ends, such as the design, manufacture, and operation of efficient and economical structures, machines, processes, and systems* [10].

Although the POC team is often composed of people who consider themselves schooled in the hard sciences, the activity is engineering. This is true even though the primary work may not be performed by someone with an engineering degree.

In addition, there are newly created engineering disciplines such as genetic and biological engineering, where engineer and geneticist or biologist are one and the same. For example, *genetic engineering* refers to the applied techniques of genetics and biotechnology that are used to cut and join together genetic material, especially DNA. These strands of DNA, which originate from one or more species, are spliced together and then introduced into an organism in order to change one or more of its characteristics [11].

While genetic engineering may be done in connection with scientific discovery, it is an engineering activity.

Biological engineering is similar in its concept and definition. According to the MIT Department of Biological Engineering Web site, "[The purpose of the] biological engineering discipline is to advance fundamental understanding of how biological systems operate and to develop effective biology-based technologies for applications across a wide spectrum of societal needs including breakthroughs in diagnosis, treatment, and prevention of disease, in design of novel materials, devices, and processes, and in enhancing environmental health" [12].

By some definitions, an engineer is described as someone who is limited to designing or building machines. However, it is obvious that in today's environment, the activity of engineering encompasses a much wider range of activities and skills.

This merging of scientific and mathematical principles into practical ends is not a new concept. In fact, it was envisioned by William Barton Rogers when he founded MIT in 1861. The words used to found MIT were mind and hand, *mens et manus*; it is the literal coupling of science and engineering [13].

At the time of MIT's founding, the classical definition of engineering was the work of applying known laws of science to the development of something. However, in the modern age many devices are developed concurrent with the development of the laws of physics that govern them. Semiconductor engineering is a classic example.

Visit a newly built semiconductor facility and you will find process and electrical engineers working together with device physicists, surface chemists, statisticians, and information technologists as yield engineering teams. A yield team for a microelectronic mechanical system (MEMS) semiconductor process will often be researching a new physical phenomenon that is adversely affecting the process. This generally involves designing sophisticated experiments to discover the mechanism causing the dysfunction.

For example, the problem could be that single layers of unwanted molecules are attracted to important submicrometer-size surfaces by tiny electrostatic forces causing an unwanted signal (e.g., interference). Solving this problem involves basic research into molecular transport and surface binding.

Is this the work of a scientist or an engineer? The answer is, both! At its heart, it is an engineering activity accomplished by an interdisciplinary team of scientists and engineers where the line between these disciplines is blurred.

To complicate further our understanding of the engineer and what engineering is, many of the premier institutions of higher education offer engineering sciences as a discipline. Formal training in engineering science includes core physical science instruction along with training in traditional engineering courses.

The course description of engineering science at the University of California at Berkeley, for instance, is intended to provide students with dual citizenship in engineering and science. It follows the core curriculum of engineering methods while enabling students to pursue interests in the areas of natural science. Thus, engineering sciences are intended to produce individuals with the capabilities to create the equipment necessary to pursue basic research within their selected area of natural science.

There is a direct correlation between this academic milieu and the world inhabited by the early physicists who researched the world of quantum mechanics. Our brief stories in quantum physics in the late 1800s in the prior chapter showed a blurring of the strict boundaries between physical sciences and engineering.

In the context of our argument regarding concurrent engineering, what is important is not the formal training of the individuals but their knowledge set and the activities they performed.

Concurrent Engineering and Biology, 1850–1880

Although it is true that in most contexts today there exists a separation between the engineer and biologist, it was not always so. The period of 1850 to about 1880 marked an era of significant advancement in biology, especially when compared to prior centuries. As such, it makes an apt example of the power of concurrent engineering.

The following brief stories make this point. They also demonstrate the subtle shift from biologists who used engineering skills to create new technology to more specialized biologists relying on technology invented and engineered by others. The biologist after 1900 is more like the observing astronomer that Dr. Mountain described in the previous chapter as opposed to the astrophysicist like himself and Dr. Doxsey who help developed the new tools to enable their science.

Starting with Pasteur, 1862

Louis Pasteur's elegant experiment to demonstrate that bacteria, carried on airborne particles, grew in broth was convincing and dramatic.

Pasteur was a chemist with a keen understanding and interest in microbial life. At the time of his work, there was considerable debate between scientists who believed in spontaneous generation and those who believed in the biogenesis/germ theory. In essence, one side of the debate thought that life could spontaneously emerge while the other argued that existing microscopic organisms could explain a number of occurrences such as development of disease, food spoilage, and more.

Pasteur was not the first to postulate the existence of biogenesis and germ theory, but he found the means to prove these theories to be correct. His experiment centered on demonstrating that airborne particles containing microbes were the cause of spoilage in chicken broth. To do this, he devised glassware specifically designed to control the admission of microorganisms carried by airborne particles to the broth.

One set of glassware had a top with a long, thin, S-shaped tube at the top, now known as a swan's neck flask. The tube would allow air to enter the flask but trap airborne particles that could not overcome the gravitational forces around the S turn. The other set of glassware essentially had an open top that allowed particles to reach the broth.

Pasteur heated the two flasks in order to sterilize the broth, and then he allowed them to sit in the air. The glassware that was opened to the air formed microbial growth while the flask protected by the S-shaped cap remained clear.

The implication of this finding was that microscopic germs could cause everything from food spoilage to disease, which in turn led to a finer understanding of the need for sterilization. It also led to the expansion of microbiology as a formal area of study within biology.

At the core of Pasteur's discovery was an ingenious experimental apparatus that was simple in its design but elegant in its engineering.

The story of Pasteur also helps us measure the time between the creation of the engineered technology and the biological discovery.

When helping develop the thesis of this book, Dr. Jeffrey Drazen (editor in chief of the *New England Journal of Medicine* and physician at the Brigham and Women's Hospital) offered the observation that from the researcher's point of view, "The clock starts when you advance a hypothesis that accounts for an observed or predicted phenomenon whose elemental basis remains obscure. You stop the clock when you do an experiment that proves the hypothesis you have offered is not untrue." What this book attempts to point out is that there is another clock that starts ticking when what becomes an enabling technology is first developed or shown to be possible in a POC. A major milestone is reached when the technology is made available to the biology community. The clock is stopped when the technology is finally used for discovery. Throughout this book, we start the clock at the moment when the science or first principles of the technology are demonstrated (POC), and we stop the clock at the moment the theory underlying the biological discovery is proved.

As an example of concurrent engineering, the gap between POC and discovery in Pasteur's story is zero.

Kirchhoff and Bunsen (1860) and Hoppe-Seyler (1861)

In 1859, physicist Gustav Kirchhoff and chemist Robert Bunsen were working together in Heidelberg, Germany, to try to better understand the properties of different chemical elements. By burning them with a Bunsen burner and then reflecting that light through a prism, they noticed that each element emitted a slightly different spectrum of light. This spectrum could then act as a unique signature for that element and help them identify slight distinctions between the different compounds.

To help them better separate the emitted light into its constituent color spectrum, the two men invented the first spectrometer. With the spectrometer they could more easily catalog the various spectrographs made from different elements and engage in a much deeper form of spectral analysis.

This then allowed them to discover cesium and rubidium and detect the presence of sodium within the sun. This latter discovery occurred when they compared the spectra emitted by the sun and the unique spectral signature made by sodium.

In all, Kirchhoff and Bunsen are credited with initiating the science of spectral analysis and received considerable renown when they published their first paper on their findings in 1860.

Of course, the story does not end here. Around the same time and about 150 km from Heidelberg, Felix Hoppe-Seyler was appointed as the professor of chemistry at a medical facility in Tubingen, Germany. He had received his medical doctorate in 1851 in Berlin but found that he much preferred to continue his studies and research in physiologic chemistry rather than practice medicine.

Quite remarkably, it was only a year after Kirchhoff and Bunsen published their findings that Hoppe-Seyler published his own analysis of hemoglobin using the Kirchhoff and Bunsen spectral analysis techniques. He was able to show spectroscopically that oxygen is bound to hemoglobin in the blood and that it could easily be replaced by carbon monoxide.

By demonstrating that hemoglobin is the oxygen carrier in the blood, he was able to bring about a fundamental understanding of the critical biochemistry of human blood.

In terms of the topic of this book, the distance in time between the POC for Kirchhoff and Bunsen's spectrograph and Hoppe-Seyler's discovery was only several months. Although Hoppe-Seyler did not discover the spectrograph, his story demonstrates the power of quickly placing the right tool in the right hands.

However, as we shall see, the distance between biologist and engineer is beginning to grow.

The Evolution of Engineering and Biological Research, 1880 to Today

After this chapter, we will begin to tell the story of how the divergence of engineering and biology has in essence created the need for concurrent engineering. It is not a story of fault or failure, merely one that demonstrates the evolution of biological research. Researchers in biology became specialists and observers and left the world of engineering behind. Rather than rely on their own engineering prowess, they became dependent upon the marketplace to deliver necessary technology.

Over the past 100 or more years, medical science has made incredible breakthroughs. The medicine that was available to the average person at the beginning of the past century is barely recognizable to that available to the modern world. However, there are still many pressing needs. When examined within the context of concurrent engineering, the paradigm that has dominated the past 100 years in biological discovery appears plodding or ambling.

To help us identify the key biology breakthroughs for the stories contained in the following chapters, we have relied on the committee that awards Nobel Prizes in this area of science. By looking at Nobel laureates and their discoveries, we have access to a consistent population of breakthrough research that is well documented.

The down side, of course, is that it represents a subpopulation of research as it does not include such areas of endeavor as advancement in epidemiology. As epidemiologist Hans-Olav Adami [14], a member of the Nobel Prize selection committee, said, "Epidemiology has a hard time meeting the criterion of groundbreaking discovery, sharply defined in time, and assignable to no more than three individuals."

It could also be argued that the Nobel Prizes in Physiology or Medicine represent a population where discovery tends to be technology intensive. Even if that is the case, it will not change the findings of this book as we dissect the engineering and technology that supported the discoveries.

With the Nobel Prize as our guide and with our metrics established, the results are fascinating. The research shows that time after time, the biological advancements that led to a Nobel Prize followed the availability of new enabling technology by a mere 5 to 10 years. However, this time gap was preceded by a much longer delay of 20 to 50 years between the POC for the technology and when it was transformed as an apparatus that scientists could use and made available in the marketplace.

What is striking about these data is the relatively long gestation time needed for the typical technology development compared to the relatively rapid speed at which researchers applied the technology. As we will show in chapter 10, *technology development to commercialization and availability to the life science community studying genetics averaged 40 years. The time to discovery measured from the availability of the technology to publication of the findings was about 13 years.*

In other words, the enabling technology takes three times as long to develop as the life science discovery enabled by it. This has startling implications.

Before the traditional biology research community can have access to a given technology, the technology has to have gone through a successful commercial development process. This process never begins unless a company detects a viable market for the product.

This in turn raises a sobering question: How many technologies with the potential for changing the course of life science have never seen the light of day because of market potential?

Thus there is the potential to change the paradigm of the past 130 years. By integrating the work of engineers and biologists and providing funding required to develop promising enabling technologies, we can eliminate the commercial viability hurdle. This in and of itself would greatly compress the time between technology concept and discovery. We show in chapter 10 that concurrent engineering and biology research teams can reduce this timescale from 40 years to an average of 7 years, or an 80% reduction of time.

At this point, we return to the notion of the anonymous engineer. Ask an educated layperson the role that engineering has played in the evolution of the life sciences, specifically biology, and something wholly predictable is likely to happen. He will scratch his head, reflect for a moment, and then answer a different question.

He or she is likely to point to technologies—such as imaging and endoscopy—that have represented major advances in the clinical world (e.g., doctors' offices, clinics, and hospitals) but will completely overlook advancements in the life sciences. From this perspective, the role of the engineer in the life sciences is to provide technologies that aid doctors and hospitals in the delivery of health care as opposed to enabling groundbreaking research into the cause and cure of disease.

Therefore, it is important to bring the engineer out from the dark, so to speak. This book seeks to elucidate the role engineering has played through the past 100 or more years of medical breakthroughs in life science research and discovery and to enable critical advances in our understanding of biology, disease, and cure.

Every year we applaud the winners of the Nobel Prize in Physiology or Medicine, and, of course, that applause is well deserved. However, standing behind these distinguished biologists are legions of unsung of technologists, engineers, inventors, and tinkerers who come up with the instruments that make the work of the biologists in their laboratories possible.

"Well," you might respond, "it's too bad that those anonymous engineers don't get their due. Nevertheless, the system seems to work. We seem to get the scientific breakthroughs that we need, right?"

Yes and no. The current system does well, but as the data demonstrate, it could be far more effective and efficient.

Currently, the efforts of the engineer and life scientist are effectively decoupled from each other. Typically, the engineer works to address a specific technological challenge that, if solved, represents a known market opportunity. The biologist, meanwhile, addresses a scientific puzzle that may be open-ended, with no payoff in sight other than new knowledge. He or she tends to use tools and technologies that are understandable, adaptable, available, and affordable. In addition, the modern pressures to publish in academia make it difficult for co-leadership of a research project by an engineer and a scientist as discussed in chapter 13.

Let's put the worst possible face on it: If the market doesn't call for it, the engineer doesn't invent it. If the engineer doesn't invent it, the biologist doesn't get access to it and the discovery does not occur.

Are we getting the scientific breakthroughs that we need and which might ultimately lead to improved human health and happiness? Probably not. Are we getting them as quickly as we could? Almost certainly not.

Some Important Questions to Be Answered

While this book is intended to move the conversation forward as opposed to being the last word, it will address the following important technology development questions:

What's possible? Can the convergence of engineering/technology with life sciences have the potential to reduce the total life science discovery cycle from an average of 50 years to, say, 10 years by making the technology available in advance of commercialization?

What has to happen first? Engineering has enabled research and clinical procedures/protocols that were never envisioned by the biologist because those procedures and protocols were far beyond the tools the biologists were trained on. But the effective development of new engineered technology has to be guided by a set of needs or requirements. We have a "chicken and egg" problem that has to be addressed. How important is the articulation of a scientific hypothesis to guidance of the engineering development? What guidance does the past provide on this point? How many of the innovations were "hatched" to meet one need and ended up meeting a very different one? Does the goal of cost reduction on the therapeutic end provide a vehicle for important innovation?

How can conflicting value systems be brought into sync? There is a natural tension between the intellectual/academic value system and that of a goal-oriented technology development project. Faculty need to publish intellectually important papers. But a goal-oriented project will often be best served by choosing the lowest-risk development path, which may not lend itself to premier publication. How can we balance the interests of the scientific principal investigators of the project with the desire to translate the knowledge to the bedside as quickly as possible?

Is a focus on development of technologies that can reduce the cost of analysis, diagnostics, and therapy important? Given the health care cost-containment debates now raging at the state and national level, surely the answer to this question needs to be "yes." But what does this mean for the advancement of science? What does it mean for the clinical operations that we seek to reshape?

What can we say about funding models and federal oversight? Most current models for life science funding (private and public) do not explicitly recognize that technology must be developed to enable life sciences breakthroughs. Current federal organizations are not configured to address concurrent engineering and leading-edge biology as the technology and biology expertise are neither integrated nor collocated.

What does the future hold? Based on what we've learned from the past, what can we say with confidence about the future? What unmet medical needs will be addressed by the engineering community over the next decade? Is that satisfactory?

II

FROM PEAS TO GENOME: ENGINEERING-ENABLED BIOLOGICAL RESEARCH

A number of case studies had to be chosen to expose the theme of the book. After looking though a number of review articles and other studies, the author settled on a list of discoveries that were largely taken from Eric Lander's lecture series on genetics. These discoveries were used to form the framework of chapters 5 to 10. Then the author looked at original articles and other sources of information to uncover the technologies that were used to enable these discoveries. These examples are just examples and not meant to be a comprehensive history of biological discovery. They were chosen from a lecture series because they represent a connected history of research.

4 Discovery of Chromosomes and the Submicrometer Microscope

Beginning in the late-nineteenth century, the majority of key advances in biology center on our increased understanding of the science of genetics. As such, this area of research makes for a compelling narrative with which to begin to draw the line between engineering and biology.

It is a story that begins with an earnest monk working to develop and then prove his hypothesis of genetic inheritance. In this early stage, the most sophisticated technology used was the ability of the researcher to make visual observations. Subsequent researchers managed to carry the baton forward, though they relied solely on the innovative use of traditional chemistry. This is an important science, but its research is centered on deduced implications of phenomena created within the chemist's lab.

Biologists weren't able directly to observe the mechanisms of genetics and cell function until the engineering-related challenges of microscopy were overcome and they could look inside the cell. It wasn't that microscopes were unavailable. The challenge was that they could not gain the magnification and resolution required to view chromosomes and the other microscopic constituents of genetics and cells.

Future advances in our understanding of genetics would rely on equally as important technologies created within the auspices of engineering.

Therefore, it can be said that while Gregor Mendel made the first important step in genetics, engineering enabled each subsequent stride on the path toward decoding the building blocks of life.

Mendel's Peas, 1866

Gregor Mendel's breakthroughs were so startling and central to the development of the science of genetics that he is now considered to be the father of this branch of science.

As many of us learned back in junior high school, Mendel (born in the Austrian state of Moravia in 1822) was a brilliant child whose farmer parents couldn't afford to educate him at university. Mendel became an Augustinian monk, which granted him the opportunity to continue his studies and to teach high school–aged students and live up to the promise of his intellect.

Mendel loved nature, especially plant life. On his walks around the monastery, he took note of the distinct characteristics of individual plants. For example, the flowers of common pea plants could either be purple or white, the stems could be long or short, and the color of the pods could be yellow or green, even though they were all the same type of pea plant. This led him to wonder how plants acquired those characteristics as well as whether the offspring would have the same traits or some intermediate mix.

The dominant evolutionary theory of the day was Lamarckism, espoused by the French biologist Jean-Baptiste Lamarck. He argued that individual organisms acquire new traits in response to their environment and these traits are passed down to subsequent generations. His theory represented a form of soft evolution. As Mendel began his work, most European scientists—including several who taught a young Charles Darwin—subscribed to some version of Lamarckism.

Mendel's natural curiosity drove him to test Lamarck's theories. Working first with mice, Mendel wanted to see if darker-colored mice mated with albinos would have a mix of both parents or if one color type would dominate over the other. However, monastic modesty being what it was at the time, a bishop forbade Mendel from having anything to do with sex [15].

Mendel then moved on to pea plants. In the course of several years of experiments, he discovered that Lamarck was wrong. Hereditary inputs, he concluded, are fixed. They are received unchanged from two parents with no environmental mediation whatsoever. Additionally, not all offspring of the same plants receive exactly the same genetic inheritance. Some traits are dominant over others.

This was an astounding set of insights that he made public in 1866 in a paper he titled *Experiments on Plant Hybrids*. Mendel effectively demonstrated the existence of genes—not a word he used, of course—and laid the groundwork for all genetic investigations to follow.

Unfortunately, his paper would remain mostly unread by the larger universe of scientists for the remainder of Mendel's life. It was not until after his death in 1884 that his work was rediscovered in 1900 by botanists Hugo de Vries and Carl Correns. These men were both working to understand the science behind heredity and helped make Mendel famous for his prescience in the study of genetic inheritance.

It does not diminish Mendel's accomplishment to point out that his research methods were fundamentally nontechnical. He tracked seven characteristics of pea plants that were visible to the naked eye. Darwin, too, used unaided observation to make his celebrated breakthroughs. Mendel and Darwin also lived and did their work in almost the exact same historical era. Mendel was born in 1822 and Darwin was born in 1809 and died 2 years before Mendel. They were exemplars of biologists who could make seminal breakthroughs independent of technology.

The Early Biochemists, 1830–1890

Prior to 1830, numerous chemists became adept at using what now are considered basic tools of chemistry to analyze various substances. For example, Sir Humphry Davy was the first to chemically characterize nitrous oxide and to notice its physiologic properties as an anesthetic. Around the same time, Jean-Baptiste Dumas similarly characterized the chemical properties of chloroform. Prior to these two men, in the late-eighteenth century French chemist Antoine François, Comte de Fourcroy [16], isolated several substances, including urea from urine.

These were important discoveries in and of themselves, but they also advanced the ability of chemists to analyze the makeup of various compounds. This in turn led to a series of gateway discoveries centered on the function and constituents of proteins.

In 1838, Gerardus Johannes Mulder published a paper in which he deduced [17] by chemical methods the elemental composition of several proteins. He also extracted and isolated the amino acids (building blocks of proteins) leucine and glycine from a hydrolysate of muscle. Unfortunately, little progress was made in understanding proteins until around 1926 when James B. Sumner was the first to isolate an enzyme, urease, and then prove that this enzyme is in fact a protein. This discovery earned him the Nobel Prize in Chemistry in 1946.

In 1869, Friedrich Miescher noticed a precipitate of an unknown substance while performing experiments on the chemical composition of cells. After further experimentation, he recognized that the precipitate's properties during the isolation procedure as well as its resistance to protease digestion (breakdown of a protein into its constituents) indicated the novel substance was not a protein or lipid.

Recognizing that he was on to something important, Miescher continued his testing in Felix Hoppe-Seyler's laboratory at the University of Tubingen. Analysis of the substance's elementary composition revealed that, unlike proteins, the precipitate contained large amounts of phosphorus.

With this discovery, Miescher recognized that he had discovered a novel molecule. As he had isolated it from the cells' nuclei, he named it nuclein, which later became known as nucleic acid. In subsequent work, Miescher showed that nucleic acid is a characteristic component of all nuclei and hypothesized that it could be involved in cell mitosis, the method of cell replication.

To say the least, this was an incredibly important breakthrough. Though he did not realize it at the time, he had discovered the basis for deoxyribonucleic acid (DNA) and ribonucleic acid (RNA). The stage was set for later scientists to broaden our understanding of DNA and RNA, their constituents, and processes.

Microscopy and the Discovery of Chromosomes, 1878

Observation in nature and the chemistry laboratory had so far proved to be powerful tools capable of laying the groundwork for the burgeoning study of genetics. However, most of the understanding of nucleic acid and the functioning of cells remained to be discovered. Theories existed that nucleic acid played a role in mitosis, yet no one had managed to identify the constituents of mitosis much less see it in action.

This was to change with the development and commercial success of the Zeiss microscope.

Carl Zeiss

Carl Zeiss founded Zeiss Werke in 1856 after serving as an apprentice lens maker. Zeiss had already made a significant contribution to the world of optics. In the 1840s, he had developed a set of lenses with an aperture range that greatly increased the resolution of images viewed through a microscope. However, he knew this would not be enough fully to distinguish his company, so he began to search for someone who could help him improve the quality and performance of his microscopes.

At the University of Jena, Zeiss came across a young physicist named Ernst Abbe. Zeiss was immediately taken by Abbe's disciplined and theoretically grounded approach to lens design and hired him in 1866 as a scientific consultant. With the addition of Abbe, Zeiss' company moved away from the traditional *cut-and-try* method of lens making to one in favor of a more rigorous approach based on physics and mathematics.

This new methodology led to a number of design breakthroughs such as a more efficient eyepiece as well as a device that improved lighting. However, the most important innovation was the *Abbe sine condition.*

In practical terms, these technical improvements meant that Zeiss and Abbe could create microscope lenses with a very high quality of resolution, which would enable even greater magnification.

In 1876, Abbe and Zeiss brought to market the first microscope capable of overcoming centuries of resolution limitations. Their microscope presented a clear image of elements as small as one-millionth of a meter, or 1 micrometer (1 μm). It was an instant sensation. The sciences of chemistry, physics, and biology were blossoming at this time, which meant there was significant demand for high-quality microscopes in Europe and abroad. Zeiss sold 1,237 microscopes between 1856 and 1870. In the first year of the new high-performance microscope's release, the company sold 556 [18] of them.

One of these was to a man named Walther Flemming.

Seeing Is Believing

Two years after Mendel published his seminal work and concurrent with the work of Miescher, Walther Flemming, a German physician turned anatomy professor, began digging into the intricacies of cell division and replication. As with Miescher, Flemming's work on this process led him to the hypothesis that the cell nucleus was important to mitosis.

Flemming knew he needed actually to observe the role of the components in the nucleus in order to understand fully the process of how cells divide and replicate. The two challenges in his way were finding a microscope powerful enough to see down to the 1-μm level and the means to illuminate the structures within the nucleus.

The first of these was overcome by access to a Zeiss microscope. The second was overcome by Flemming's innovative use of newly developed aniline dyes to highlight structures in the nucleus without washing out the entire view of the cell.

In 1878, Flemming was the first to identify chromosomes and to outline the process of mitosis while observing slides of cells from salamander larva [19]. He did not see the process of division and replication as it occurred—the cells were killed by the dye—but he was able to see cells in various stages as they underwent mitosis.

This was an enormously important discovery. It uncovered the structure on which genes reside and eventually opened the door to decades of exploration on the function of these molecules in the life cycles of cells.

Timeline of Proof of Concept to Breakthrough

When examining Flemming's story within the context of concurrent engineering, we see there was a modest gap between the proof of concept and Flemming's breakthrough. Abbe developed the first principles that underlie the Abbe sine condition in 1872 [20]. The proof of concept was achieved with the first working models of the microscope in 1876. Then in 1878, Flemming achieved his breakthrough using the Abbe–Zeiss microscope.

Though the gap is small when compared to those of later years, it highlights an important point. Flemming and his peers could neither unlock the mysteries of cell division and replication nor begin the study of chromosomes until the underlying technology of the microscope had been improved. As it turns out, development of a microscope that provides visibility down to the single-micrometer level played a critical role in what remains a monumental biological discovery.

Pioneers of Light Spectroscopy

Concurrent (though not part of the Lander lectures) with the breakthroughs in the nascent study of chromosomes and genes, scientists were starting to take a closer look at proteins. To the non-biologist, proteins are understood to be the structural components of the body (tissue, hair, nails, muscles, blood, etc.).

While proteins can self-assemble to build structures, they also act as the mediators for the messaging that controls cell growth. The challenge for the protein researcher in the late-nineteenth century was that proteins were too small to see with even the Zeiss microscope.

Once again, Ernst Abbe's insights inform our understanding. He proposed that understanding of light wavelengths represents not just a means to increase the effectiveness of microscopy but also an aid to define the outer limits of resolution.

This meant that new approaches were needed for characterizing important biological molecules. This, of course, would change in the mid-twentieth century with the advent of the electron microscope. In the interim, scientists devised other clever technologies to probe the composition of the cell.

David Alter, Anders Ångström, Gustav Kirchhoff, and Robert Bunsen, among others, explored the interactions of various materials with light. In 1854, David Alter characterized the spectra from different metals vaporized by an electric spark. Ångström added the theory that the absorption and emission wavelengths of light from incandescent gases are the same. Kirchhoff codified Ångström's observations into three laws related to light emitted and absorbed from a hot object. Bunsen went on to detail the spectra from various metals heated to incandescence.

By 1885, there were a number of scientists developing a theoretical understanding of spectral emissions. Johann Balmer was one of them. He was able to identify a predictable interval between the four measurable emission wavelengths of hydrogen and thus opened the door to understanding the quantum nature of the atom.

As numerous scientists used spectral analysis to study materials, there was little published on how to apply the technology to understand biological materials better. As in most of biology, researchers had to wait for a new technology—absorption spectra, in this instance—to become available commercially. One of the earliest instruments accessible to biologists was the Michelson Echelon Diffraction Grating and Spectroscope.

Absorption Measurements of Proteins and Amino Acids, 1900–1930

In July 1901, Adam Hilger, who owned an optics manufacturing company (figure 4.1), published a paper on the method for properly using the Michelson Echelon Diffraction Grating, a particularly powerful spectroscope sold by his company. In that publication, he included an advertisement for his own optical equipment. The advertisement, shown in figure 4.2, appeared in the January 1907 issue of the journal *Nature*.

Prior to 1901, physicists used spectrographs to understand various chemical elements better and, in the case of Balmer, to create hypotheses as to the structure and nature of the atom. One of the first biology-related uses was by the physiologic chemist Felix Hoppe-Seyler. He used the Kirchhoff–Bunsen techniques in 1861 to discover that oxygen is bound to hemoglobin and is easily replaced by carbon monoxide.

While Hoppe-Seyler clearly has a biological connection, it took the widespread commercial availability of spectrographic equipment, such as by Hilger, for biologists to overcome the limits of microscopy. Once the equipment was available and the necessary expertise in its use developed, discoveries, especially in the research of proteins, soon followed.

In 1915, Philip Adolph Kober published the findings of his spectrographic studies of amino acids, the building blocks of proteins [21]. In 1928, Frank Campbell Smith carried this work forward by examining the absorption spectra of uric acid [22]. Both scientists used equipment manufactured and sold by Hilger.

Though these two researchers carried the study of proteins forward, they faced certain limitations. For example, the Hilger instruments were constrained by bulk properties—such as light absorption—of the protein molecules being studied. Essentially, visibility was limited to the broad outlines of the nanometer-sized molecules they were studying. Structural details of these molecules weren't worked out for another 50 years until the development of technologies able to see down to the nanometer level.

ADAM HILGER,
75a, CAMDEN ROAD,
LONDON, N.W.

MAKER OF ALL KINDS OF

HIGH-CLASS OPTICAL INSTRUMENTS,

INCLUDING—

Table Spectroscopes and Spectrographs.
Goniometers.
Solar and Stellar Spectroscopes and Spectrographs.
Solar Protuberance Spectroscopes.
(J. EVERSHED'S DESIGN.
Pocket and other Direct Vision Spectroscopes.

MICROMETERS of all kinds,

Including PHOTOMEASURING MICROMETERS, with High-Quality Screws.

COELOSTATS AND HELIOSTATS.

PRISMS OF GLASS, QUARTZ, ICELAND SPAR, FLUOR SPAR, ROCKSALT, or any other material.
A LARGE STOCK of QUARTZ and ICELAND SPAR always in hand.

PLANE PARALLEL GLASS A SPECIALITY.

PREPARATIONS of ICELAND SPAR and QUARTZ for the study of POLARISED LIGHT, including:—

Nicol, Foucault and Hartnack Prisms; Rochon and Wollaston Double Image Prisms in Iceland Spar or Quartz; Savart Plates, etc., etc., etc.

PRICE LIST ON APPLICATION.

Figure 4.1
Adam Hilger advertisement of optical equipment.
Source: Instruments of Science 1800–1914, the Michelson Echelon Diffraction Grating 1901.

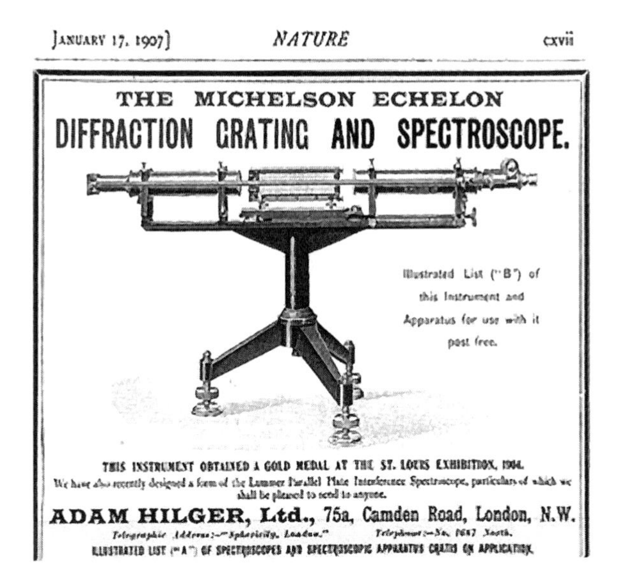

Figure 4.2
Adam Hilger advertisement for the Michelson Echelon Diffraction Grating.
Source: Nature, January 17, 1907.

The next significant advancement in spectrography came with the development of ultraviolet light–based spectrometers. In 1915, S. Judd Lewis led the way in the development and manufacture of this type of spectrometer in order to enable his work of measuring the absorption spectra of a number of serum proteins. The Lewis spectrometer was first offered in an article in 1917 [23].

Although the advances of Kober, Smith, and Lewis are not in and of themselves monumentally significant in terms of our understanding of genetics, they were important first steps. They laid the groundwork not just for the discoveries that were to follow but also for developing the technical expertise among biologists to work with the relatively new and evolving technology of spectrography.

They, with Flemming and others, also represent a very good example of the interdependence between biology and engineering. Their work and discoveries were deeply dependent upon technological advances created under the auspices of engineering.

5 DNA: Gels, Paper, and Columns

Progress in the study of genetics slowed somewhat in the early twentieth century, but it does mark an important period of technological advancement that led to a wealth of important discoveries.

Like Mendel, Thomas Hunt Morgan and his student Alfred Sturtevant were able to uncover the fundamental mechanism in the control of inherited traits without the aid of newly developed technology. The basis of their experiment involved crossing multiple variants of fruit flies and then observing the traits inherited by subsequent generations.

In 1914, Sturtevant created a theoretical genetic map from the likely positions of genes on the chromosomes based on the outcome of multiple crosses of flies with different traits. Morgan, for his part, used the experiments in his "Fly Room" at Columbia University to show that chromosomes carry genes and that genes are the fundamental mechanism for heredity [24]. This in turn led him to be awarded the Nobel Prize in Physiology or Medicine in 1933.

About fourteen years later (1928), Frederick Griffith, like Sturtevant, made a major leap in biology with creative experimentation independent of new technology. His experiments uncovered a transforming principle upon which genetic traits are carried.

Griffith used mice to test the virulence of two strains of *Streptococcus pneumoniae*: one strain (S) was lethal to mice while the other one (R) was benign. He showed that while the R strain as well as heat-inactivated bacteria of the S strain were not deadly by themselves, when combined they were inevitably lethal to inoculated mice. He concluded that the live R bacteria could not only acquire the virulence trait from the heat-treated, dead S bacteria but also propagate that trait to future generations. As a sad postscript, Griffith and an assistant were killed by a German bomb while working in his laboratory in 1941.

These discoveries in the first 30 years of the twentieth century were important and clearly set the stage for dramatic and rapid breakthroughs to come.

However, these future breakthroughs depended upon the development of technologies such as ultracentrifugation, gel electrophoresis, and chromatography.

Gel Electrophoresis and the Rockefeller Foundation

If we think of the early study of genetics as the hunt for the components of inheritance and cell function, we can look at it as a winnowing process. First came the understanding that all cells contain a nucleus and that the nucleus plays an important role. This was followed by the discovery that the nucleus contains chromosomes that themselves hold genes, the mechanism of genetics.

Further research led scientists to find that genes are segments contained within DNA strands, which hold four chemical bases: adenine (A), guanine (G), cytosine (C), and thymine (T). These bases form the alphabet of genetics, and when placed in specific orderings they spell out the instructions for how a cell is to function. Within this complex interplay is the role and function of proteins.

To say the least, the above represents a gross simplification of the components of genetics. Further, these discoveries were not made in a precise chronological order. What is important to note, though, is that identifying and characterizing the components of genetics required researchers to break cells down into their constituent parts. They then had to identify each part and deduce an understanding of their role and function.

This is not simple science and required the development of ingenious technologies by people with highly evolved engineering skills.

Finding the Transforming Principle

Frederick Griffith may have discovered the existence of a transforming principle, but in 1928 he did not have the means to identify the actual molecule that caused the transformation. This important breakthrough would not occur until 1943 when researchers at the Rockefeller Institute for Medical Research began using a relatively new set of technologies.

Oswald Avery, Colin MacLeod, and Maclyn McCarty at the Rockefeller Institute for Medical Research sought to uncover the actual molecule in the Griffith experiment that caused the transformation. To achieve this, they used known enzymes to create different subproducts from the lethal S strain and then separate the constituents in the newly developed gel electrophoresis apparatus, which was developed by a fellow researcher at the Rockefeller Institute.

They separated the components of S strain lysate using a sequence of enzymatic removal until they were left with the active portion that could confer virulence to the benign R strain. "Examination of the same active preparation was carried out by electrophoresis in the Tiselius apparatus and revealed only a single electrophoretic component of relatively high mobility comparable to that of a nucleic acid." This led them to conclude, "The evidence presented supports the belief that a nucleic acid of the deoxyribose type is the fundamental unit of the transforming principle of Pneumococcus Type III" [25].

The Avery, MacLeod, and McCarty work was published in 1943. At that time, the use of gel electrophoresis was just coming on the scene, as the first reliable apparatus for the technique was developed and built at the Rockefeller Institute from 1936 to 1939.

Gel Electrophoresis and Ultracentrifugation

Although electrophoresis and ultracentrifugation matured between 1920 and 1940, the early work of creating methods to separate molecules began around 1809. This is when Ferdinand Ruess published the results of his experiments on the migration of colloidal (e.g., molecules held in suspension) clay particles under the influence of an applied electric field [26].

Then in 1892, Harold Picton and S.E. Linder reported the movement of ions while under the influence of an electric field. In fact, they were probably the first to observe ionic deposition on a cathode (e.g., particles held in suspension where an electric field is applied and the particles migrate to one pole). This early work by Picton and Linder represents the formative stages of the development of electrophoresis.

About this same time, the Swedish physical chemist Theodor Svedberg was engaged in his work to understand better the chemistry and physics of colloidal dispersion systems [27]. In particular, he wanted to determine some of the properties of proteins.

To conduct this research, he needed a means to separate proteins from other cellular and background components. This led him to invent the ultracentrifuge in 1921. Svedberg's ultracentrifuge in principle is similar to centrifuges found in laboratories back then as well as now.

The ultracentrifuge causes the various particles held in suspension to separate and form sedimentary bands. These colloidal dispersions are essentially the different particles segregated by their shape and mass.

For his work on dispersal systems, Svedberg was awarded the Nobel Prize in Chemistry in 1926.

This, of course, was not the zenith of the ultracentrifuge's useful life. Physicists Jesse Beams and Edward Pickels of the University of Virginia furthered the development of the ultracentrifuge in 1940. Subsequently, Pickels set up a company called Spinco (Specialized Instruments Corporation) to commercialize the ultracentrifuge and sold more than 300 of them.

At nearly the same time as Svedberg, Beams and Pickels worked on the ultracentrifuge, Arne Tiselius was perfecting gel electrophoresis. A former student and devotee of Svedberg, Tiselius initially struggled with difficulties related to the thermal control and diffusion of his first electrophoresis devices.

Much as it sounds, the primary medium of gel electrophoresis is the gel. Most modern electrophoresis is conducted through an agarose gel first developed by Nobel Laureate Phillip Sharp (interview later in this book), which is a semi-firm gelatin. The gel is formed into a rectangle about the size of a piece of paper and of the order an inch thick or less (modern gels are a fraction of an inch thick). At one end, slits are cut out to form wells. The molecules suspended in a liquid solution are pipetted into the wells.

The gel is placed in a container with a constant electrical field across the gel. The samples, such as DNA and protein, are pipetted into the wells, and when the electrical field is applied, they are drawn through the gel toward the opposite pole.

Because the interior of the gel is a porous mesh, the smaller molecules travel more quickly through the gel than the larger molecules. For example, DNA strands vary in length so over a set period of time the longer strands will not travel as far across the gel as the shorter strands. Strands that are the same length end up grouped together and form bands. These bands can be seen when they are stained with special dyes.

As Tiselius was perfecting his early version of electrophoresis, the electric potential required to cause the molecules to migrate created an electrical current in the gel, which caused localized heating. This heating disrupted the uniform conditions required for gel electrophoresis to work properly and distorted the bands of separated molecules.

In 1934, Tiselius renewed his efforts to improve electrophoresis after talking with protein researchers at Princeton. All of them articulated the unmet need for a protein separation and characterization technology. With funding from the Rockefeller Foundation Fellowship, he created a new, large apparatus in 1936 which eliminated the distortion from heating. This new device allowed for observation of the separation as well as the removal of the different fractions (e.g., sorted groups of molecules) for further study.

By 1939, there were 14 machines in use in the United States, and five of them were at the Rockefeller Institute laboratories where Avery and his team had access to them. In 1945, the American manufacturer Klett Manufacturing Company made the first commercial electrophoresis machine. By 1950, four different companies were making them, including Spinco and PerkinElmer.

Eventually, Spinco was acquired by Beckman Coulter and moved into chromatography as well. As a side note, William Dreyer, the noted California Institute of Technology biology professor, consulted for Spinco and played a significant role in developing an automated protein sequencer with them. He was also the mentor to Leroy Hood [28], who is credited with developing an automated DNA sequencer.

For his work on electrophoresis and adsorption analysis, Arne Tiselius was awarded the Nobel Prize in Chemistry in 1948.

Increased Use of Gel Electrophoresis

Before the widespread availability of gel electrophoresis, protein research advancement was carried forth via sequential separation and extraction methods. However, these processes were complicated, time consuming, and imperfect.

For example, the biochemist William Rose [29] isolated and characterized the last-identified amino acid, threonine, in 1935. As he describes the process of isolation, "12 kilos of commercial fibrin were hydrolyzed with sulfuric acid and, after removing the acid, was concentrated by vacuum and filtered, treated with copper carbonate to form copper salts of the amino acids, and extraction with alcohol. Purity was achieved when a crystal formed." He then used optical reflectance angle measurement as well as freezing point depression and melting point measurements to distinguish one amino acid from another.

When combined with the work of others, Rose's addition completed the catalog of the 20 standard amino acids. All of these were identified using quantitative fractionation (an imperfect separation process) and the formation of insoluble salts of acids.

In the mid-1930s, James B. Sumner went on to characterize in greater detail the nature of proteins, including the proof that enzymes are proteins [30]. His work was done largely with the chemical crystallization method [31] described earlier. In his Nobel lecture he laments, "At this time I had no access to the ultra-centrifuge of Svedberg or to the electrophoresis apparatus of Tiselius."

Why didn't he have access to Tiselius's technology? Tiselius first published his work in 1930, and Sumner was exploring the structure of proteins in 1936 (the same year Tiselius completed his prototype). Although Tiselius's first publication was in 1930, the device was not widely available. It was not

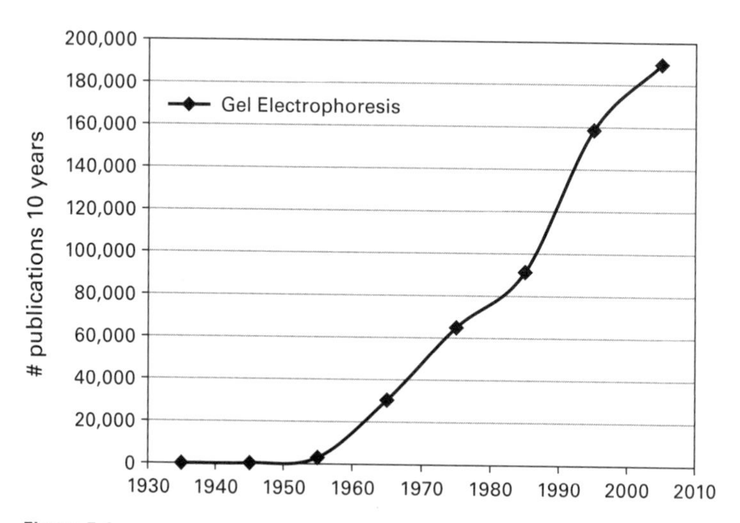

Figure 5.1
Scientific articles involving gel electrophoresis.

until 1938 that researchers had broad access to a standard version of the technology.

As noted earlier, upon broader availability of the Tiselius apparatus, the use of the technology blossomed. Between 1930 and 1940, there were a mere 30 publications in which the technology was used. By contrast, electrophoresis has become a mainstay tool for biological research approaching 20,000 appearances in published papers each year between 2000 and 2010 (figure 5.1).

One of the key advances with the technology that led to its wider use was the development of capillary electrophoresis in the late 1960s and 1970s. By this point, improvements in electrophoresis fractionation ability were sought by many.

The first advancement came in 1967 by S. Hjerton, which was followed in 1974 by Rauno Virtanen, and then Franciscus Edmundus Petrus Mikkers in 1978 [32]. The advantage of the capillary design was reduced thermal convection and mixing of the boundaries allowing higher voltages and much better separation efficiencies.

Further advances included the use of fluorescent dyes to mark the various targets and create better resolution.

The Chargaff Rule and Chromatography

Even though gel electrophoresis was an incredibly important tool for research into DNA, proteins, and other studies, it was not an end unto itself. Researchers still had to find new means and methods to answer vexing questions related to the study of genetics.

One of these very important questions was the mechanism behind the immense variety of life on the planet. If DNA is the genetic material of all living things, it must be configurable in many ways in order to explain the variation between and within species.

The Tiselius gel electrophoresis apparatus had already proved very effective at isolating molecules and allowing scientists to run experiments that address biology and genetics at the molecular level. This work ushered in the study of molecular biology.

With electrophoresis in hand, Avery and his team demonstrated that DNA is the genetic material. Upon publication of their findings in 1944, the door was opened to the biochemist Erwin Chargaff to dig into the properties of DNA. Research had already shown that the controlling mechanism was very likely centered on the four nucleic acids: adenine (A), guanine (G), cytosine (C), and thymine (T).

Chargaff believed that if he could characterize the amount of each of the four nucleic acids (AGCT) in DNA and between species, he could reveal the mechanics of genetic variation. As with Avery, a new technology was just making its way into biology laboratories that would serve Chargaff well.

Paper chromatography had just become an important research biologist's tool, but it was not necessarily a new concept to science. Generally speaking, chromatography is the collective term for a set of techniques to separate individual molecules held in mixed suspension based on various differences in their physical properties. Although chromatography uses a range of techniques, the principle separation is achieved by sorting the molecules either by size (primarily filtration), charge, attraction to certain substances, or chemical reactions. Chromatography can be used either as an analytical research method unto itself or for the preparation of pure samples of particular constituents.

The first significant instantiation of chromatography was in 1890 when the Russian botanist Mikhail Tsvet developed the first column fraction chromatography. He then demonstrated its effectiveness in 1906 by using it to separate plant pigments. Essentially, he used a long glass tube filled with an absorptive material such as calcium carbonate. He then poured a slurry of plant pigments into the tube and allowed them to percolate down through the tube. Because different pigments moved at different rates through the absorptive material, they separated into colored bands in the tube.

As with other enabling technologies examined in this book, there was no immediate commercial interest, and the technology faded.

Then in 1941, Archer John Porter Martin and Richard L.M. Synge picked up the development of the technology. While studying proteins, they were able to refine the process of paper chromatography.

Chargaff used the basics of this new technology developed by Synge and Martin but modified it to suit his needs. As he stated in his 1950 paper, "The literature is not rich in procedures that would lend themselves to a complete survey of the nucleotide constituents of nucleic acids. The most promising approach appeared to be the employment of a chromatographic method, which was indicated by the successful application of filter paper chromatography to the separation and estimation of purines and pyrimidines" [33].

Armed with this new chromatographic approach and ultraviolet light, Chargaff separated DNA into its component parts and discovered the first of his rules related to the four nucleic acids. Within the structure of DNA, the amount of adenine (A) is equal to the amount of thymine (T), and the amount of cytosine (C) equals guanine (G), otherwise known as *A = T and C = G*. This rule is known as base pairing and demonstrates the molecular construction of DNA molecules.

In 1950, Chargaff published his findings, which soon became commonly known as the *Chargaff rules*. He also shared this information with James Watson and Francis Crick, which guided their model for the structure of DNA. They realized that DNA was probably a complementary double helix structure with AT and GC as complementary base pairs.

Sanger Insulin Sequence

Chargaff was not the only researcher to use chromatography to good effect.

Biochemist and two-time Nobel Prize winner Frederick Sanger started his training in 1936 at the University of Cambridge where he studied natural sciences. He merged his training in physiology and chemistry and focused on biochemistry, which was a new department at Cambridge. He then worked under Albert Neuberger studying the metabolism of the amino acid lysine in the animal body. His training gave him insights into both biology and biochemistry, which led him to make the decision to apply his talents aggressively on understanding proteins.

As he later stated, "My main interest at the time was amino acid analysis of proteins, especially of insulin."

Sanger followed the lead of Nobel laureate Hermann Emil Fischer, who synthesized a number of amino acids and identified peptide bonds, which are the structures that link amino acids in a protein. Fischer's work depended entirely on a chemical approach to proteins, and Sanger recognized that this offered a promising pathway for his work with insulin.

Sanger knew that he needed a reagent for determining the end groups of proteins. He also was anxious to use the newly discovered technology of partition chromatography [34]. He had managed to show that the disulfide bonds within the peptides could be broken by oxidation, but he had not yet found a method to fractionate the chains of peptides. The ability to perform this type of separation and sorting was important to enable the structural analysis of insulin that Sanger was attempting.

In 1947, he visited Arne Tiselius's laboratory in Uppsala, Sweden. By that time, Tiselius was considered a leader in the field of protein research and was well known for his electrophoresis device. Upon arrival and seeing the work under way, Sanger came to realize that the chromatographic system based on charcoal adsorption being developed at the laboratory offered hope. Sanger discussed this with Tiselius, who suggested that Sanger should spend some time working in the laboratory in Uppsala [34].

Soon after, Sanger met Richard L.M. Synge, "who was working there and who introduced me to zone ionophoresis, which was the forerunner of other more successful ionophoreses such as paper or acrylamide gel."

Synge was not only working on ionophoresis at Uppsala, but also he and Martin were developing and advancing the field of chromatography. In particular, they were seeking to improve the separation capability of column chromatography by adding a countercurrent solvent extraction method. After months of trial and error, they were able to construct the first column chromatograph with silica gel [35].

However, while this type of column chromatograph worked well for numerous separations, it had limitations, especially when separating certain amino acids [36]. Synge and Martin continued to work on separation technologies and found that paper-based partition chromatography worked well for amino acids. They then went on to develop two-dimensional paper chromatography.

The development by Synge and Martin of paper-based partition chromatography was an incredibly important new technology for Sanger. He later wrote, "With the insulin chains we were again lucky in that the method of paper chromatography had just been developed by Martin and colleagues. The fractionation of small peptides was far superior to anything that had been achieved previously and it seemed something of a miracle at the time to see these hitherto intractable products separated from one and another on a simple sheet of filter paper."

Although paper chromatography and electrophoresis were critical technologies that Sanger capitalized on for work in proteins and then DNA, this was not enough. He needed to develop various chemical agents to cleave and mark the amino acids at specific locations so they could be displayed via chromatography. He developed the *Sanger reagent,* fluorodinitrobenzene (FDNB), to react with the exposed amino groups in the protein with the N-terminal amino group at one end.

He then partially broke down (hydrolyzed) the insulin into the amino acid segments, either with hydrochloric acid or by using enzymes such as trypsin. The mixture of peptides could then be displayed in two dimensions by first using electrophoresis in one dimension and then, perpendicular to that, by chromatography in the other. This was essentially a two-step separation method that created a far purer and defined end product.

The different peptide fragments of insulin created distinct patterns that Sanger called fingerprints. The peptide from the N terminus could be recognized by the yellow color imparted by the FDNB label. By repeating this type of procedure, Sanger was able to determine the amino acid sequences of peptides generated by using different methods for the initial partial hydrolysis.

These smaller sequences could then be assembled into the longer sequence to deduce the complete structure of insulin.

Expanding Use of Chromatography

Chromatography entered general laboratory use following a path different from other technologies, such as electrophoresis. Once Avery and Martin published the numerous separations they had performed between 1941 and 1948, the technology (which required only a few components) was widely sought out by the scientific community. Adoption and access to standard devices was aided by several suppliers.

The spread of the technology to the laboratories from this point on was rapid and not paced by a commercial development lead time but rather by a spread of knowledge over a 7-year span. And with expansive use came even more research breakthroughs.

Paper chromatography contributed to the determination of the structure of antibiotic peptides. Then Frederick Sanger used paper chromatography to figure out the structure of the insulin molecule. During these same years, Melvin Calvin used paper chromatography to separate components of photosynthesis. Chargaff also used this technology to determine the ratios of bases (ATCG) in DNA and hence developed the understanding of base pairing. These examples demonstrate the rate at which paper chromatography had evolved into a mainstream tool. They also show that engineering can act as a catalyst and enabler of discovery.

Still, there was nearly 40 years between the demonstration (proof of concept) of first principles and the widespread usage of paper chromatography.

Gel Electrophoresis and Chromatography

Figure 5.2 summarizes the two tracks of development of electrophoresis and chromatography. The hexagonal items are commercialization events.

Although the fundamentals of electrophoresis were known and published in 1930, it was not until 1950 when Spinco and PerkinElmer started offering electrophoresis technology on a wide scale. Additionally, nearly 40 years elapsed between the first demonstration of capillary electrophoresis by Tiselius and the first commercial unit in 1989.

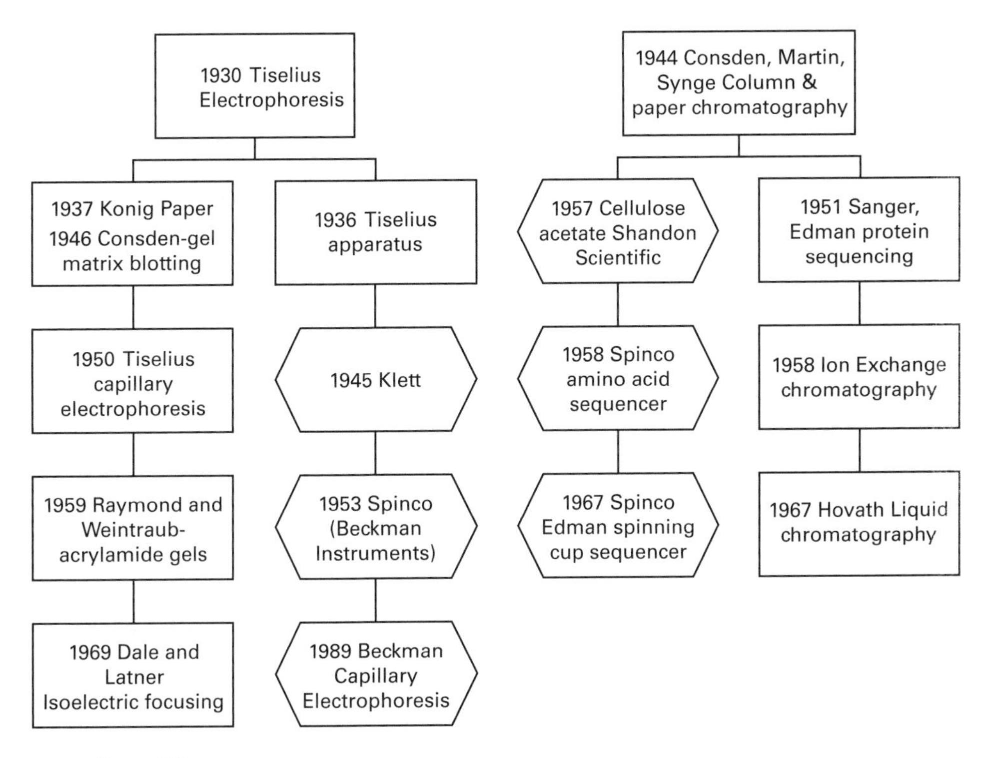

Figure 5.2
Molecular separation technologies of electrophoresis and chromatography. Developments (boxes) and commercialization (hexagons) of some (not all) important milestones in gel electrophoresis and chromatography. Both technologies rely on differential migration of molecules through a medium such as gel, paper, or package column.

Parallel to this, the number of publications with electrophoresis in the title grew with the commercialization of each of the technology advances. The initial leap in the study of genetics in 1950 was probably in part the result of the availability of the Spinco and PerkinElmer instruments. Then in 1989, the first fully automated capillary electrophoresis instrument (P/ACE 2000) was offered by Beckman, which coincides with the explosion in the number of electrophoresis articles.

A similar evolution of improvements and commercialization took place with chromatography except that the first demonstration of paper chromatography by Tsvet was in 1890. It was not until 1944 that Martin and Synge demonstrated the effectiveness of paper chromatography. The simplicity of the device allowed for the rapid spread and use of the technology rather than reliance on the pace of commercial development of an automated instrument.

6 Structure of DNA and Proteins: X-ray Diffraction

By 1943, Avery, MacLeod, and McCarty had shown in their elegant experiment that DNA is the genetic messenger. As a result, the race was on to understand how the message of life is generated and transmitted from parent to child. There were numerous teams working on the DNA mystery, but the most prominent centers of molecular biology were the California Institute of Technology (Caltech), King's College (London), and the University of Cambridge.

The research undertaken in the United Kingdom consisted of two lines of investigation. The first was the experimental work of Maurice Wilkins and Rosalind Franklin at King's College. Their work centered on developing methods to crystallize DNA and then characterize the crystalline structure via X-ray diffraction.

The second team was composed of the molecular biologists James Watson and Francis Crick at the University of Cambridge. These two theorists were attempting to put together the known DNA data into a cohesive hypothesis of the structure and function of DNA.

Prior to their famous pairing, Crick had worked in the Medical Research Council unit at the University of Cambridge, which was headed by Max Perutz and located in the university's Cavendish Laboratory. Watson, an American, arrived at Cambridge after a chance meeting with Maurice Wilkins at a symposium in Naples, Italy. As a result of this meeting, Watson developed an interest in conducting research on the structural chemistry of nucleic acids and proteins [37]. In early August 1951, arrangements were made for Watson to begin work at the Cavendish Laboratory, where he met Crick.

As the two teams of investigators in the United Kingdom gained steam, there were still members of the research community who believed that proteins were the genetic material. They maintained this belief even with the inferred results of the Avery, MacLeod, and McCarty experiment. However, Watson was convinced that DNA was the genetic material and focused his energy in this direction [38].

Wilkins and Franklin and Watson and Crick were not the only molecular biologists in pursuit of the structure and function of DNA. Several teams of molecular biologists at a number of locations were also racing to be the first to create a model of DNA and understand its behavior. Most notably were Max Delbrück and Linus Pauling at Caltech, the epicenter of this type of research in the United States.

Pauling, who was well steeped in X-ray diffraction, had just published his discovery of the alpha helix structure of protein, which the group at Cavendish tried to discover as well but were beaten by Pauling. This past competition fueled the fire to be the first to understand the structure of DNA.

Feeding the Crick and Watson work were numerous observations about the nature of DNA. The most prominent of these was the work of Rosalind Franklin, who used X-ray diffraction to create a series of high-quality images of DNA. As many familiar with this story will know, Wilkins showed one of these photos (famously known as Photo 51), taken by Franklin by X-ray diffraction, to Watson in January 1953 without the consent of Franklin. This has spurred decades of controversy over the role of Franklin and the credit she deserves.

Upon seeing the photo, Watson almost immediately recognized the double helix structure of DNA. Aiding this finding was Chargaff's observations about the relative composition of the basic building blocks of DNA, namely the four bases: adenine (A), guanine (G), cytosine (C), and thymine (T).

Watson and Crick had the confirmation they needed to publish their results in the April edition of the journal *Nature* under the title "A Structure for Deoxyribose Nucleic Acid."

In 1962, the pair along with Maurice Wilkins was awarded the Nobel Prize in Physiology or Medicine.

Watson, Crick, and the Double Helix: Putting It All Together

It is hard to capture the insights of Watson and Crick in a few sentences. However, our focus is not on their work as much as it is to identify the technology and engineering that led to their discovery of the structure of DNA as a double helix.

As outlined in chapter 5, Erwin Chargaff had observed the remarkable relationship between the base pairs, where the quantity of A and T is always equal (A = T) and the quantity of C and G is always equal (C = G). Watson and Crick's insight was that the A = T and C = G rule must be reflected in the structure of DNA.

They also knew from Rosalind Franklin's work that DNA is a helix and that there are sugar–phosphate chains—the framework of the helix—located on the outside of the molecule. This latter finding gave critical insight as to the structure of the helix as well as to its molecular stability.

With insights gained from Chargaff and Franklin, Watson and Crick constructed the graceful intertwining double helix structure of DNA. Their findings and model were supported by and complied with key experimental observations.

The discovery of the double helix represents the single biggest leap in understanding in modern molecular biology. The story of Watson and Crick is known by nearly every schoolchild studying biology, and their findings have formed the basis of whole areas of medical research and breakthroughs.

Within their story, X-ray diffraction is the key technology that enabled the conceptual model of the double helix. The history of X-ray diffraction will be discussed in a moment. It is important to note for now that without this research tool, neither Watson and Crick nor any of the other scientists in the race to model DNA could have succeeded.

The story of X-ray diffraction's contributions to biological research does not end here. Around the same time, Max Perutz, also at the University of Cambridge, was studying the structure of hemoglobin with X-ray diffraction.

Hemoglobin and Max Perutz

In 1936, Max Perutz began his work to create a model of the structure of hemoglobin at the Cavendish Laboratory as a research student under John Desmond Bernal [39]. At the time, Perutz was supported financially by his father and was lucky enough to have landed with Bernal, who was considered a leader in the use of X-ray diffraction in molecular biology.

With the annexation of Austria and Czechoslovakia by Hitler in 1938, Perutz lost the financial resources of his father. However, he was able to continue at Cavendish after being appointed in 1939 as an assistant to Sir William Lawrence Bragg, director of the Cavendish Laboratory [39]. Again, Perutz was lucky to have landed with a leader in the use of X-ray diffraction. Bragg was awarded the Nobel Prize in Physics in 1915 for developing Bragg's law of X-ray diffraction, which is the basis for determining the crystalline structure of molecules.

While still active in this line of research, Perutz was appointed in 1947 to head the Medical Research Council (MRC) Unit for the Study of the Molecular Structure of Biological Systems. In 1962, this laboratory was moved to a new location and renamed the Medical Research Council Laboratory of Molecular Biology, or LMB as it is commonly known. During its 1947 instantiation, the laboratory had a rather august set of members that included

Frederick Sanger (1951), Francis Crick, and James Watson (1950), among other notables.

It was a dynamic environment for molecular biology, and Max Perutz was at the heart of things. Similar to the story of the structure of DNA, Perutz was not the only researcher in the hunt for the molecular structure of hemoglobin. Linus Pauling was also focused on the discovery of the fundamental structure of hemoglobin as he worked to elucidate his general theory of protein structure. Over a number of years, Pauling's research had led him to a series of findings that culminated in 1951 with the structural model of protein that Pauling called the alpha helix.

"Max was thunderstruck. He could see at once that the structure must be right" [40] wrote Georgina Ferry in her biography of Perutz. Despite being thunderstruck, Perutz went on to confirm experimentally a critical repeat dimension of the Pauling model and gave Pauling the result, for which he was grateful. After helping confirm the alpha helix configuration as a protein building block, Perutz sought to apply this approach to his own work on the large protein molecule of hemoglobin.

As his research continued, Perutz's work on the structure of hemoglobin and Crick's work at the MRC unit were intertwined [40]. In several unit meetings, the structure of hemoglobin was discussed, and Crick was a major contributor to the genesis of the model that Perutz arrived at.

Perutz was a superb experimentalist, but he still faced huge technical and financial challenges. Undaunted, Perutz and his team worked diligently on a significant technical problem of X-ray diffraction holding up their progress. They needed to improve the signal and information they were deriving from the diffraction patterns. They had seen recent work that suggested that if they introduce a heavy atom like mercury into the hemoglobin structure at specific locations, they could enhance the analysis of the complex array of diffraction patterns.

Crick confirmed theoretically that the idea would work, and Perutz went about the development process to create it. He hired Vernon Ingram, a biologist with considerable technical skill, who was recommended to Perutz to help develop the heavy atom process.

When they finally succeeded in labeling the molecule and looked at the diffraction patterns, "Max knew that his eureka moment had arrived" [40].

In 1959, Perutz reported that he had discovered the molecular structure of the protein hemoglobin. Then in 1962, he was awarded the Nobel Prize in Chemistry, which he shared with laboratory partner John Kendrew, who discovered the molecular structure of myoglobin through the use of X-ray diffraction.

Throughout his search for the structure of hemoglobin, Perutz and his team demonstrated an ability to develop necessary technology as they moved forward. This included the heavy atom technology as well as technology to improve the diffractometer itself. They also developed the computational capacity to derive images from the many thousands of diffraction patterns they created.

Ultimately, the meaning of Perutz's work to science was not just revealing the structure and function of hemoglobin but also advancing the capabilities of X-ray diffraction. The success of Perutz's technology had a huge impact on subsequent elucidations of protein structure and function.

X-ray Diffraction

The story of X-ray diffraction begins back in 1910. Max von Laue working with Walter Friedrich and Paul Knipping exposed copper sulfate crystals to X-rays and recorded the resultant diffraction pattern on a photographic plate. In 1912, Sir William Lawrence Bragg demonstrated that the diffraction patterns of X-rays followed specific relations that could be used to determine the size and spacing of the atoms in the crystalline material.

A few years later, Philips, an electronics company in Eindhoven, The Netherlands, developed an interest in this new technology and began exploratory work on the functional use of X-rays. In the mid-1920s, Philips put an engineering team together and developed a safe X-ray tube, which initiated the development of X-ray as a medical and industrial device.

Still, progress toward a reliable X-ray diffractometer was slow. Finally, in 1945 the U.S. Naval Research Laboratory sponsored work at North American Philips to develop the world's first commercial X-ray diffractometer, which was initially produced under the Norelco brand (figure 6.1).

From the moment of its first instantiation to its introduction as a commercial device in 1945, physicists and chemists relied on purpose-built X-ray diffraction technology of their own creation. For instance, Bragg, Bernal, and Dorothy Crowfoot Hodgkin were leading crystallographers of the time who developed the equipment they needed as they needed it.

With the advent of a reliable commercial product, marketing of the diffractometer grew and by 1950 it was a globally available instrument. A wider group of researchers could apply the technology to the questions they wished to investigate without having to develop the instrument themselves. Enter Rosalind Franklin and Maurice Wilkins at King's College.

Franklin was intent on characterizing DNA with X-ray diffraction. After working on crystallizing DNA and through a myriad of trials, Franklin was able to get a definitive X-ray diffraction pattern of DNA, which is the now

VOLUME 25, NO. 5, MAY 1953 53 A

Will X-Ray Diffraction Reveal the Hidden Secrets

of Cancer . . . Heart Diseases . . . Hardening of Arteries?

Serving Science and Industry

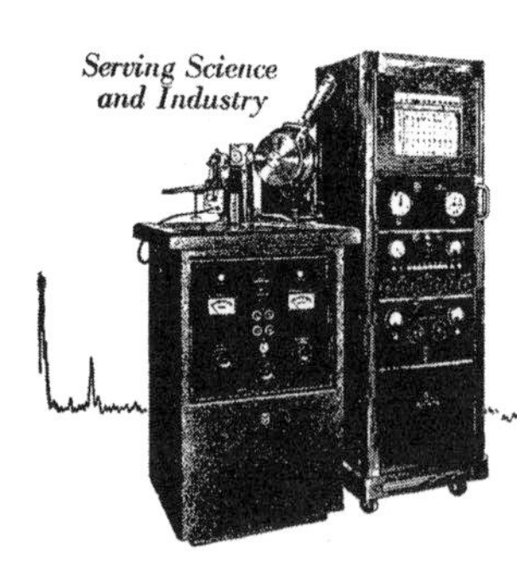

DOCTORS at New York's Mount Sinai Hospital are using NORELCO X-ray Diffraction equipment designed by Philips—world leader in X-ray development – to observe molecular behavior in the calcium salts of bone cancer, calcium salt deposits in artery walls, and the composition and formation of kidney stones. New observations have been made concerning heart structure and behavior. In X-ray Diffraction, there is indeed new hope.

Applied to industry, NORELCO X-ray Diffraction equipment has already revolutionized metallurgical and chemical methods – improved controls for production, processing and end-product analysis.

In the scientific laboratory, it has permitted non-destructive analysis impossible for traditional chemistry.

You may find NORELCO X-ray Diffraction equipment just as valuable for your company or research organization. Write for free consultation. Also free booklet—"What is X-ray Diffraction?"

A Norelco PRODUCT OF NORTH AMERICAN PHILIPS COMPANY, INC.

Dept. IE-5 • 100 East 42nd St., New York 17, N. Y.

In Canada: Rogers Majestic Electronics Ltd., 11-19 Brentcliffe Road, Leaside, Toronto 17, Ontario

Figure 6.1
Norelco (Philips) X-ray diffractometer developed in 1945 in cooperation with the U.S. Naval Research Laboratory.
Source: www.panalytical.com. Reprinted with permission from *Analytical Chemistry*, May 1, 1953. Copyright 1953 American Chemical Society.

famous Photo 51. As noted earlier, Wilkins showed this photograph to Watson without Franklin's permission or knowledge on January 30, 1953, when Watson visited King's College.

In March 1953, Crick and Watson completed their model of DNA with Photo 51 as a critical support of their concept. From the time that the first diffractometer was commercially available (around 1947) to the publishing of the double helix (April 1953), there is a 6-year gap. However, it took 35 years from the proof of concept of the first principles of X-ray diffraction to the commercial development of the enabling technology (1912 to 1947).

X-ray Diffraction at the Laboratory of Molecular Research

Although the structure of DNA had been identified, work continued at the Laboratory of Molecular Research on the structure of proteins to discover their nature and behavior.

In 1956, Walter J. Kauzmann [41] developed an understanding of protein folding, which is the process by which a chain of amino acids (polypeptides) transitions into a formed state such as an alpha helix. His work was mostly theoretical, but it provides an interesting insight into the introduction of technology to support medical research.

Even as late as 1959, the use of X-ray diffraction for protein structure had not been widely deployed. In his classic paper [42], Kauzmann summarized the technologies available for observing changes in the configurations of polypeptide chains in proteins. Among the several optical observation methodologies, he included X-ray diffraction.

"If a protein can be obtained in crystalline form, the wide angle X-ray diffraction pattern is potentially the most promising method of obtaining detailed information on the structure," he wrote. "The field is undergoing active development and progress has been reviewed repeatedly (see, for instance Kendrew and Perutz, 1957)" [42].

As of 1959, the work to map the structure of a protein by X-ray diffraction had not yet been completed. This was true even though the structure of DNA by X-ray diffraction had been performed by Rosalind Franklin 8 years earlier. However, Max Perutz was just about to publish the structure of hemoglobin by X-ray diffraction and change the protein landscape.

Applying X-ray diffraction to protein crystals, in 1959 Perutz was able to determine the molecular structure of the protein hemoglobin. Earlier work had provided researchers with information on the general shape of protein. However, it wasn't until X-ray diffraction came into use that researchers had a powerful enough tool to determine the interatomic distances and bond angles. These are critical measures that illuminate structure.

As noted earlier, Perutz began his work on hemoglobin using X-ray diffraction in 1939 at the Cavendish Laboratory of the University of Cambridge under Sir William Lawrence Bragg. By 1953, they had successfully developed an approach to measure the structure of the protein.

However, as important as all of these data were, they did not clearly spell out the molecular model of hemoglobin. The team had volumes of data and complicated calculations that had to be performed, which led to reliance on another relatively new technology. As Perutz notes in his Nobel lecture, the number of computations required to discern the message spelled out by the spots in each X-ray diffraction image have been overwhelming.

"Clearly, this would have been impossible before the advent of high-speed computers," he said, "and we have in fact been very fortunate, because the (local) development of computers has always just kept in step with the expanding needs of our X-ray analysis."

Helping Perutz was Hilary Muirhead (identified by Perutz in his Nobel lecture). She performed all of the programming of the complex results of the diffraction patterns to resolve the location of the heavy atom labels in the hemoglobin molecule [43]. These were markers added to the hemoglobin at specific locations to aid measurement and analysis of the diffraction images. Muirhead performed the programming on a computer developed at the University of Cambridge called EDSAC.

The Electronic Delay Storage Automatic Calculator (EDSAC) was developed in 1949 by Maurice Wilkes, a computer science engineer and director of the Mathematics Laboratory at Cambridge. His goal was to develop a practical computer that could be used by any faculty at the university. The model he relied on to develop his computer was the EDVAC design by John von Neumann, a giant in the world of mathematics and computational theory.

Von Neumann's key conceptual design contribution revolved around a computer architecture where data and the software program are stored within the computer.

With von Neumann's insights in mind, Wilkes successfully developed the EDSAC, one of the first computers with an internally stored program. Later, he developed a concept for microprogramming the central processing unit (CPU) with an application-specific program in read-only memory (ROM). These two modifications greatly increased the processing speed of computers. This latter design element was incorporated into the EDSAC II, commissioned in 1958 (much of this history can be found in "IEEE Annals of the History of Computing," Vol. 14, Issue 4, October 1992). The EDSAC II played a prominent role in John Kendrew's work to model myoglobin based on X-ray diffraction patterns.

These fascinating computers used mercury delay line architecture for memory. The information in the form of electrical pulses was transduced in mechanical waves in mercury-filled tubes (figure 6.2). When the signal returned to the start of the tube having reflected off at the end of the tube, the pulses were re-introduced, thus refreshing the memory (the pulse).

In addition to the technique of adding heavy atoms to specific sites of the hemoglobin molecule and the use of the first storage electronic computers, Perutz and his team also:

1. engineered rotating anode tubes (patented by the team) for the angstrom-level (one ten-billionth of a meter) resolution required;
2. designed and built a microdensitometer, which measured spectral density;
3. developed Fourier transform algorithms for processing the data.

In all, Perutz names 20 team members in his Nobel lecture with a sentence or two on each of them and their contribution. In recent tributes, Perutz was recognized for his remarkable ability to identify, attract, and inspire great

Figure 6.2
Maurice Wilkes shown with a 16 delay line memory tank unit.
Source: Andrew Herbert, The EDSAC Replica Project (http://www.bscmsrc.eu/sites/default/files/media/andrew_barcelona-oct2011.pdf).

minds from different disciplines in his 23-year pursuit to characterize fully the structure of proteins.

Throughout his later life, Perutz kept his scientific focus and over two decades amassed multidisciplinary teams capable of building the technology he needed to achieve his goals. Perutz also gave birth to an organization called the Laboratory of Molecular Biology at Cambridge that has produced nine Nobel Prizes shared among 13 Nobel laureates. This is a remarkable achievement and will be discussed further.

Although Perutz's fertile mind and impressive organizational skills were important to his work, his findings owe no small amount of debt to technology developed under the auspices of engineering.

7 Observing DNA and Protein in Action: Radioisotope Labels

Avery and his team as well as Watson, Crick, and Perutz had performed yeoman's work to deduce the nature and structure of DNA and proteins via the use of technology. In some cases, they produced their own enabling technologies, while in others they made use of research tools as they became commercially available through companies such as Spinco, Philips, and others.

The next wave of discovery unleashed by these advances would be similarly reliant on largely anonymous engineers. The mystery of structure had been overcome, but function presented a range of new challenges. In particular, researchers needed a means to see or deduce the processes of genetic transfer from one generation to the next as well as the method of DNA replication.

The tool that came to the fore was the use of radioisotope labels, which allowed researchers essentially to monitor the processes of DNA and proteins remotely. The first to effectively use this new technology, however, were two researchers seeking to confirm that DNA is in fact the genetic material.

Hershey–Chase Experiment, 1952

In 1952, Alfred Hershey and Martha Chase conducted a series of experiments that confirmed the work of Avery, MacLeod, and McCarty that DNA is the genetic material. To tell the story properly, a little context is important.

The work of Hershey and Chase slightly predates that of Watson and Crick as well as the famous Photo 51 by Rosalind Franklin. It also followed Avery and his team by about 9 years. As with Avery, Hershey and Chase labored against the belief by some researchers that proteins were the genetic material. Even though the experiment of Avery and colleagues was a powerful demonstration of the role of DNA, their findings were not universally accepted. Hershey and Chase set out to prove unequivocally that Avery and his team were right.

The basis of Avery's work had been to identify the source of the transforming principle by demonstrating that a DNA digesting enzyme, one that essentially destroys DNA, inhibits the inheritance of the virulence factor in *Streptococcus pneumoniae*. Hershey and Chase moved in a similar direction. In particular, they believed that bacteriophages were the perfect subject to demonstrate the method of genetic transfer.

Bacteriophages are virus-like particles that infect bacteria by injecting into the bacteria the genetic material that they carry inside an outer protein shell.

Hershey and Chase also realized that the perfect tool for their work was the newly available radioisotope labels being produced and distributed by the Oak Ridge National Laboratory [44].

As they knew that the bacteriophage was made up of a protein shell with DNA inside, they could label each of these components with a unique radioactive tracer. They labeled bacteriophages with isotopes of phosphorus and sulfur. These two elements were not selected at random. Rather, Hershey and Chase took advantage of the fact that phosphorus is in both DNA and proteins, whereas sulfur is exclusively in proteins.

The labeled bacteriophages were allowed to infect bacteria, and the progeny of the bacteria were analyzed. Hershey and Chase found that the phosphorus label was carried to the next generation, but there was no sign of the protein tracer. Ergo, it was in fact the DNA that acted as the transforming principle.

For their work, Hershey was awarded the 1969 Nobel Prize in Physiology or Medicine, which he shared with Max Delbrück and Salvador Luria, who also studied genetic transfer in viruses.

Meselson–Stahl Experiment, 1958

With the structure of DNA resolved and its role as the genetic material fully proved, researchers began work to deduce its function. As biochemist Arthur Kornberg noted, Watson and Crick established the frontier and now "high adventures of enzymology lay ahead."

In particular, he was referring to research that would unravel the biosynthesis (e.g., creation of components of DNA strands) and the process whereby individual helixes of DNA replicate themselves. Kornberg's contributions will be discussed in a moment, but first the author will look at the work of Matthew Meselson and Franklin Stahl.

Prior to Meselson and Stahl, Watson and Crick averred that DNA replication occurs through a semiconservative method. However, this concept had not yet been proved, and there were still the alternative possibilities that DNA followed either a conservative or a dispersive method of replication.

Meselson and Stahl set out to prove or disprove the hypothesis of Watson and Crick.

Often referred to as the most beautiful experiment in biology, Meselson and Stahl followed a similar path as Hershey and Chase. Meselson and Stahl used a heavy isotope of nitrogen as a tracer, which was made available to the biologists by the U.S. Atomic Energy Commission.

They started out by growing *Escherichia coli* bacteria in the presence of the heavy ^{15}N isotope of nitrogen. The cells were then switched to media with normal nitrogen (^{14}N) for precisely one round of replication. The resulting bacteria contained DNA with density that was exactly intermediate to both isotopes—thus demonstrating that the newly synthesized DNA represented one of the strands, while the other, the old strand, was directly inherited from the parent. This experiment conclusively disproved the "conservative model," which predicted that upon replication, bacteria should have either the ^{15}N or ^{14}N isotope but not a mix of the two.

It is important to note that their experiment was made possible by the commercialization of isotope labeling technology, which we will discuss in a moment. Meselson and Stahl also benefited from another key technology: the ultracentrifuge manufactured by Spinco [45].

In their 1958 article in the journal *Proceedings of the National Academy of Sciences of the United States of America* (PNAS) announcing their findings, the two researchers wrote, "We anticipate that a label which imparts to the DNA an increased density might permit an analysis of this distribution by sedimentation techniques" [46].

By the end of their experiments, they were able to conclude, "The results of the present experiment are in exact accord with the expectations of the Watson-Crick model for DNA duplication."

Not only had they proved an important function of DNA, but also they had developed a significant methodology using critical enabling technologies. In a 2004 article published in PNAS, Philip C. Hanawalt noted the lasting impact of their methodology for separating newly synthesized DNA: "It has become a classic approach for the biochemical detection of DNA strand exchange in recombination" [45].

Arthur Kornberg: DNA Replication, 1958

A biochemist, Arthur Kornberg has the somewhat unique status of playing an important role in discoveries surrounding DNA without having worked at either Caltech or the University of Cambridge. In 1953, Kornberg accepted a position as head of the Department of Microbiology at Washington University in St. Louis. It was here that he performed the work that would lead to his Nobel Prize, though he left the department in 1959 but continued to add important discoveries to the study of DNA and microbiology.

From the moment that Kornberg arrived at St. Louis, he committed himself to seeking the enzymes that are the fundamental mechanisms for the creation of DNA. As with Meselson and Stahl, he used *Escherichia coli* bacteria to seek the enzyme at the heart of DNA biosynthesis and the process of replication.

In 1956, Kornberg isolated DNA polymerase—the enzyme responsible for synthesizing DNA—using radioisotope labeling and chromatography. He then wanted to demonstrate the process of DNA replication.

His conviction was that the activated 5′ side of the polymer (a length of DNA) was the starting point for DNA replication. In his experiment, he grew DNA in a test tube for a short period of time in the presence of a radiolabeled nucleotide (cytosine, in this case). After stopping the process, he added two enzymes—one that cleaves DNA at the 5′ end and one that cleaves DNA at the 3′ end. Only the 3′ end–specific enzyme released a radiolabeled cytosine, implicating that during DNA synthesis the chain grew from the 5′ to the 3′ end.

These and other important findings—especially the multiple roles of DNA polymerase—were published in the May 1958 issue of the *Journal of Biological Chemistry*. In 1959, Kornberg was awarded the Nobel Prize in Physiology or Medicine for this work.

Radioactive Tracers

As should be clear, the Hershey–Chase, Meselson–Stahl, and Kornberg experiments and discoveries were all enabled by isotope technology that stemmed from operations overseen by the Atomic Energy Commission. However, these stories do not represent the limits of the use of this technology. Rather, they mark its beginning as an important tool in biological research.

In the mid-1950s, DNA replication was a major focus of biological research. In some respects, the door was opened by Hershey and Chase in their first experiments using radioisotope technology, but the research began in earnest with Arthur Kornberg's pioneering work. His breakthroughs were followed only a couple of years later by Meselson and Stahl in 1958.

The next step in the progression was carried out by another Kornberg, Arthur's son Roger. Arthur discovered the enzyme DNA polymerase and the replication process, while Roger added to the knowledge by characterizing the function and structure of DNA chromatin (the structure that creates the tightly wound chromosomes).

Roger then went on to unravel the mysteries of the process of transcription via RNA polymerase, which is the first step of gene expression.

The Kornberg story is particularly interesting because of the father–son relationship as well as their intense use of newly available technology to forge their path and phenomenal achievements. In his 1959 lecture for his Nobel Prize, Arthur Kornberg foreshadowed the Nobel Prize of his son by saying, "What we have learned from our studies over the past five years and what I shall present is that the replication of DNA can be examined and at least partially understood at the enzymatic level even though the secret of how DNA directs protein synthesis is still locked in the cell."

In 2006, Roger was awarded the Nobel Prize in Chemistry for unlocking the secret of transcription via RNA polymerase.

Although technology played a key role for both men, its use for Arthur began in 1955 when he started with concepts elucidated in Watson and Crick's "masterful hypothesis on the structure of DNA" [47]. From this he sought the model of replication by looking for an enzymatic process.

Arthur needed to isolate and purify the active polymerase, which he did with diethylaminoethyl cellulose and phosphocellulose column chromatography. The principles of chromatography, as discussed earlier in this book, were initially demonstrated in 1902, but it took decades to create a commercial product. Only by 1945 was there commercial chromatography technology available.

Kornberg thought he had isolated the DNA polymerase, but he needed to prove it was indeed sufficient for DNA replication. His approach was to use the newly developed radioisotope technology to label with radioactive ^{14}C and ^{32}P the A, C, T, and G bases of DNA strands held in solution. He then successfully performed DNA synthesis in a tube using the DNA polymerase–containing fraction that he had isolated via chromatography. His further insight that DNA replication occurred from the 5′ to the 3′ end was achieved by digesting the synthesized DNA with enzymes that cleaved preferentially the 5′ or the 3′ ends, as described earlier.

By 1958, he was able to say, "The conclusion from these several experiments thus seems inescapable that the base composition is replicated in the enzymatic synthesis and that hydrogen-bonding of adenine to thymine and guanine to cytosine is the guiding mechanism."

A profound breakthrough supported and enabled by technology developed prior to his findings over a number of years. As noted, paper chromatography for separation was available in the late 1940s, but how did the radioactive tracer become available?

Some of the properties of radioactive material had been understood as far back as the 1800s and had become the focus of British chemist and physicist Ernest Rutherford around the beginning the twentieth century. He encouraged his student George de Hevesy to work on isolating radioisotopes of radium in 1911. De Hevesy demonstrated the potential usefulness of radioactive "tracers" in biology in 1923, when he successfully measured the uptake and distribution of radioactive lead in plants.

A decade and a half later, radioisotope technology—that is, isotopes and detectors like Geiger counters—began to find its way into biology laboratories. However, the turning point for key subsequent developments in the field came in 1946, when the Oak Ridge National Laboratory began producing and selling medical-grade isotopes in large quantities for research and medical treatments.

The Isotope Program and related services at Oak Ridge operated as a commercial venture. It still does so today (as the National Isotope Development Center), making isotopes available to more than 40,000 medical professionals. Oak Ridge is far from a traditional business, and its isotope distribution efforts, along with those of the Brookhaven and Los Alamos National Laboratories, are pretty rarified distribution channels.

Once again, brilliant scientists can't do what they do best unless technologists and a commercializing entity set the table for them.

Radiolabeling as Precursor to Protein and DNA Sequencing

As time passed, radioisotopes became widely used tools that were ultimately applied to sequencing protein and DNA structures. Frederick Sanger helped to develop these technologies (discussed later in this book) after a visit in 1954 from Christian B. Anfinsen. A biochemist who would later win the Nobel Prize in Chemistry in 1972, Anfinsen was very enthusiastic about the new technique of labeling with radioactive isotopes.

As Sanger later said of the visit, "Previously I had assumed that isotopes were in the realm of the physicists and that the apparatus and techniques would be beyond my means. I learned that a number of radioactive substrates were available. A powerful tool when used in conjunction with paper fractionation" [34].

Sanger explored the use of radioisotopes with César Milstein to label the active center of a protein enzyme and determine its sequence. Later, Sanger used this technique to sequence DNA. Both were hugely significant breakthroughs and demonstrate why Sanger is a two-time Nobel laureate.

8 Transcription and Electron Microscopy

Thus far, the detailed mechanisms of DNA had largely remained beyond the view of the biologist in his laboratory. Without some Jules Verne–like machine to transmogrify researchers to the size of the molecules, a new form of microscopy had to be found. The challenge was to overcome the limitations of the traditional light microscope and to see beyond the cellular level (1,000 nm) to the molecular level (10 nm).

Here again, the anonymous engineer enters the fray with the development of electron microscopy aided by advancements in sample preparation by talented and inventive microbiologists.

By 1960, conclusive proof had been provided that genetic messages are carried by DNA and not proteins. Moreover, the structures of DNA and many proteins as well as the method of DNA replication had largely been identified. What remained was discovering how the genetic code carried by a gene within DNA expressed itself to create a gene product such as a protein.

On its surface, the process of gene expression follows a relatively simple two-phase pathway. During *transcription,* messenger RNA is synthesized from DNA so that the genetic message in a gene is copied to the RNA. Once RNA is synthesized, it is transported from the nucleus to the cytoplasm of the cell. The next phase is *translation,* where the RNA is used as a template by ribosomes to synthesize chains of amino acids held together by peptide bonds (a specific chemical bond) to form a polypeptide (e.g., protein). Finally, the unformed strand of protein goes through a process of protein folding to assume a three-dimensional structure.

Of course, in practice the above pathway is an intensely complicated process with multiple substeps and processes. Decoding how all of it works took the efforts of multiple researchers working with a range of tools, most notably electron microscopy.

Prior to electron microscopy's role in developing a working understanding of gene expression, especially transcription, there were several Nobel Prizes awarded for discovery of important individual pieces of the puzzle. For example, there is the interesting case of Severo Ochoa. He was awarded the Nobel Prize in Physiology or Medicine in 1959 ostensibly for the discovery of the enzyme responsible for synthesizing RNA from DNA during transcription.

This discovery was enabled by ion chromatography, which separates ions and molecules by their charge. However, after further analysis it was found that the enzyme polynucleotide phosphorylase (PNPase) he had isolated did not synthesize RNA but rather was involved in the degradation of unwanted RNA.

Instead, it was Samuel Weiss, Audrey Stevens, and Jerard Hurwitz who in 1960 isolated RNA polymerase while working in separate research groups. They relied on radiolabeling to analyze the components they had separated by paper electrophoresis [48].

Using this and other discoveries as well as electron microscopy, Roger Kornberg—Arthur's son—was able to reveal the actual process of transcription.

Roger Kornberg, Nucleosome, and Polymerase

His father, Arthur Kornberg, had researched DNA replication, and Roger picked up the baton by going after the next step in the genetic dogma of DNA: *transcription*. Prior to this research focus, Roger was already well known for his work at the Laboratory of Molecular Biology (LMB) in Cambridge, England, using X-ray diffraction and electron microscopy to delineate the structure of chromatin [49]. He reported his findings in 1974 [49].

There were at the time many pieces of the puzzle emerging. Component pieces to the puzzle were known. Amino acid protein building blocks were known but were assembled by an unknown mechanism into proteins. Roger's initial work on chromatin found that small segments of DNA are wound up with histones (proteins) into a tight configuration called a nucleosome. Strings of nucleosomes are wrapped tightly together by linker DNA and histones to create long strands of chromatin. With further protein scaffolding, tightly coiled chromatin strands form the X-like structure of a chromosome.

Roger trained with Aaron Klug at Cambridge where he developed his understanding of X-ray diffraction and the newly evolving technology of electron microscopy. With these tools Roger was able to add significant detail to his description of the structure of chromatin. This revealed the nature of the chromosome and the mechanism for exposing genes that are to be transcribed into proteins, which established the first of many steps of transcription.

In a published interview of Roger, he was asked whether his work was biology or chemistry [50]. He replied that, "The boundaries have become blurred between chemistry and biology." To understand the basic principles governing the biological mechanisms, he said, you have to understand the underlying chemistry, which means understanding the structural nature of the molecules and their behavior; thus, the apparent dissonance of receiving the Nobel Prize in Chemistry in 2006 for work performed within the Department of Structural Biology at Stanford University School of Medicine.

Although his work is something of a hybrid of chemistry and biology, Roger believed that chemistry is the "queen of the sciences" and at the heart of the matter. However, the fusion of the two disciplines created operational challenges, not the least of which was that to conduct his 30 years of research on transcription (chromatin and RNA polymerase), Roger had to develop creative funding strategies as no single grant would cover the entire scope of his research stretching from detailed structural chemistry to elucidation of biological processes.

One of the strategies he speaks about is the challenge of defining interim objectives, which can attract funding. One can then do related work and draw upon the funds to explore the basic science. It is understood that one can use the money for other purposes, but the stated objective must be successfully completed. This approach, of course, limits any work in new technologies to support the research to a sliver of the funding pie.

After his work at the LMB, Roger began his work at Stanford where he went on to study the details of RNA polymerase from a structural chemistry viewpoint. There are 30,000 atoms in the RNA polymerase, and Kornberg's laboratory was able to characterize how the polymerase worked and the function of many of the parts of the huge molecule.

They found how the RNA polymerase trapped the DNA and locked onto it. They were also able to create an image of the polymerase molecule in action. Supporting all of his efforts was X-ray diffraction as well as the newly developed technology of electron microscopy.

In June 2001, Roger and coauthors Patrick Cramer and David A. Bushnell published their findings in the journal *Science*. In 2006, Roger was awarded the Nobel Prize in Chemistry.

Electron Microscope

As noted, Ernst Abbe originally proposed that the ability to resolve detail in an object under magnification was limited by the wavelength of the light used to illuminate the image. By the time of Abbe's death in 1905, the magnification limits of visible (400 to 800 nm) light microscopes had been reached. In 1904, August Karl Johann Valentin Köhler, and Moritz von Rohr developed a shorter-wavelength (300 to 400 nm) ultraviolet (UV) microscope, which allowed for an increase in resolving power of about a factor of two. However, this required more expensive quartz optical components, due to the absorption of UV by glass, and ran into the limitations of light wavelength outlined by Abbe.

At this point it was believed that obtaining an image with submicrometer information was simply impossible due to this wavelength constraint. The implications for biological research, especially for molecular biologists, were obvious. Researchers were forced to rely on other means—chromatography, electrophoresis, and so forth—to deduce the structure and function of DNA, protein, and other cell components.

However, the death of Abbe and the limitations of UV optical microscopes did not mean the end of advancements in microscopy. If light is too long to enable higher resolution, a source of detection with shorter wavelengths could prove to be a major breakthrough.

That breakthrough came in 1924 when Louis de Broglie articulated the relationship of wavelength and Planck's constant divided by the particle momentum. Using de Broglie's insights, the predicted wavelength of a beam of electrons accelerated by a high voltage was found to be 0.01 nm—5,000 times smaller than that of light—and hence could provide better resolution. In other words, the limits of a light-based microscope could be overcome by energized electrons.

With de Broglie providing the theoretical underpinnings, it was now up to individuals working within the auspices of engineering to take the concept and turn it into a tangible tool.

An important antecedent technology was developed in 1858 by Julius Plücker when he recognized that the deflection of cathode rays (electrons) was possible by the use of magnetic fields [51]. This finding by Plücker paralleled the work of William Crookes, who developed the Crookes tube discussed earlier in this book. In 1897, the effect noticed by Plücker was used by Ferdinand Braun to build a primitive cathode ray oscilloscope (CRO), a measurement device.

The connection of this foundational work to microscopy was made in 1926 when Hans Busch showed that Abbe's wavelength findings could, under appropriate assumptions, be applicable to electrons.

In 1928, at the Technological University of Berlin, Max Knoll formed a team of researchers to advance the cathode ray oscilloscope design. The team consisted of several Ph.D. students including Ernst Ruska and Bodo von Borries. They concerned themselves with lens design as well as the development of electron optical components—such as magnetic lenses—which could be used to generate low-magnification images (figure 8.1).

In 1931, the group successfully generated magnified images of mesh grids placed over the anode aperture. Arguably, they had created the first electron microscope. Ruska and Knoll continued to improve upon their technology, which attracted the attention of the manufacturing company Siemens. The company became convinced that the electron microscope had enormous potential and embarked on the development of a commercial unit.

Siemens produced the first commercial transmission electron microscope (TEM) in 1939, but it had significant limitations, not the least of which was that the sample had to be less than 100 nm thick. Prior to the Siemens TEM, Manfred von Ardenne, the prolific German physicist and inventor, built the first scanning electron microscope (SEM) in 1937, thus allowing the imaging of the surface of an object of any thickness.

However, TEM and SEM technology would languish in the laboratory until it was refined and developed by Sir Charles Oatley and Gary Stewart, both of the University of Cambridge. Their work was noticed by the Cambridge Instrument Company (CIC), and in 1965 the first commercial device was marketed, nearly 30 years after Siemens's first attempt.

The SEM by CIC was called the Stereoscan, and the very first one to be manufactured was delivered to DuPont. In the first year of production (1965), CIC built five units, and sales over the next few years grew to about 100 per year.

Among those early adopters of SEM technology was the Laboratory of Molecular Biology, which put the technology to good use. In 1972, Roger Kornberg arrived at the LMB to perform his postdoctoral work.

Roger was an early adopter of the SEM at the LMB, not waiting for it to arrive at Stanford. Within 6 years of the SEM being commercially available, Kornberg used it for his initial work to understand the structure of chromatin, which was published in 1974.

Once again, development and commercialization of key enabling technology, the TEM and SEM, took nearly 34 years, and once in the market an additional 10 years elapsed until the important DNA-related breakthrough took place. After elucidating the structure of chromatin, Roger used the tools of X-ray diffraction and SEM to characterize the function and structure of RNA polymerase, which was the first critical step toward understanding transcription.

Since its development, electron microscopy has become a principal tool leading to the development of entirely new areas of study, which in turn have led to extensive breakthroughs. One can only imagine the degree to which

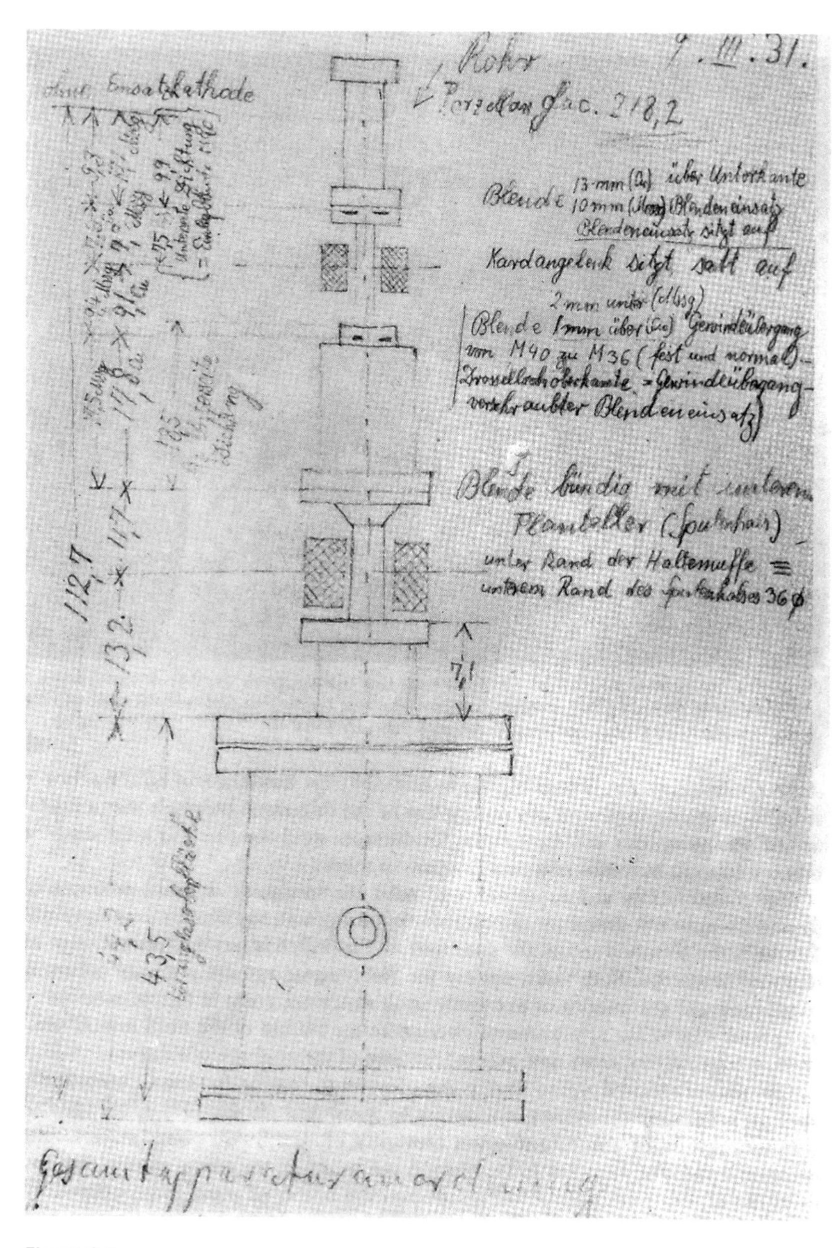

Figure 8.1
From Ernst Ruska's notebook in 1931, a sketch of the first electron microscope, originally capable of only ×16 magnification.

research could have been accelerated had this tool more easily found its way to widespread use. As such, it demonstrates the potential of concurrent engineering to speed the rate of discovery.

Phillip A. Sharp and RNA Splicing, 1977

The story of Phillip A. Sharp (Nobel Prize laureate) and his discovery of RNA splicing provides an apt illustration of not just the partnership between biology and engineering but also the intersection of physical chemistry and biology.

Trained as a physical chemist, Sharp transitioned his research focus from the domain of physical chemistry to the world of biology after he became interested in the study of DNA. Armed with the tools of electron microscopy and gel electrophoresis, he turned his newly found focus to achieve significant biological discoveries.

As Sharp began his work, initially as a chemist, the process of RNA transcription had largely been discovered. However, there were important questions that remained unanswered. One of these related to understanding how individual genetic messages (e.g., genes) are translated into proteins. Sharp was an early adopter of electron microscopy for genetic studies.

Phil has been an advocate of integrating new technology into biological research and agreed to an interview. What Phil said in the interview was a surprise. As Phil says about the author's preconceived notion that he was an accomplished biologist who learned about enabling technology along the way—"total inversion." He was an accomplished technologist who learned about biology along the way!

The following are Phil's own words:

> I got my Ph.D. at the University of Illinois, and the subject of my thesis was the hydrodynamics of worm-like chains, for which I was trained as a physical chemist. [Author's note: The understanding of these chains in a liquid was important later in Phil's work for the handling and preparation of DNA for electron microscopy analysis.] Worm-like chains is a description of how DNA behaves in solution in hydrodynamic flow; it is a very extended molecule and makes solutions very viscous. The extension of polymer statistics and hydrodynamics to DNA emerged as an interesting subject as DNA became more interesting after the discovery by Watson and Crick.

Like other Nobel Prize winners, Sharp's early years were in a physical science in which he developed analytical and investigative expertise that he applied to biological discovery later in his career. Clearly an accomplished biologist and possessing great biological insights to have achieved his status in biology, Sharp nonetheless points to his physical chemistry training as a critical underpinning of his later work.

I got interested in the question of how physical chemists were able to contribute to molecular biology by using physical techniques to understand the properties of DNA. Some of the important questions in 1968 were how long is the DNA that constitutes a chromosome and what is the topology of the DNA? Is a chromosome many different pieces or single segment? What is the structure of a chromosome? How could you map genes on chromosomes? We knew genetically where genes were, but we didn't know physically where genes were.

All of these questions were there in the literature and as a chemist there were obvious chemical contributions to be made to this field. Norman Davidson's laboratory at Caltech, more explicitly the graduate research of Ron Davis, has just developed methods to anneal two genomes and visualize deletions by electron microscopy. This was the first physical mapping of a genome and suggested that similar methods could be used to investigate the chromosomes of more complex organisms.

However, the assemblage of the bases that may have been sequenced does not give the biologist the vital information about the location of the genes on the chromosome.

These were the experiments I read. I was eager to join Norman's type of science so I applied to his laboratory at Caltech and when I arrived, Ron Davis was still there. He had just finished his first paper on Lambda deletions [a reactive end of bacteriophage DNA] and was completing a study mapping the location of RNA polymerase on the T7 phage genome.

So I went to Caltech to be with Norm Davidson, and Ron Davis trained me in electron microscopy. We overlapped about six months.

The research was done on a Philips transmission microscope, and they had an electron microscope suite with three microscopes. Microscopy has always been an important tool in biological research because "seeing is believing." Seeing things at the level of a gene was new and exciting.

Phil continues to describe his evolution to being a biologist through the exposure to Ron Davis and Norm Davidson. In addition to developing his skills with electron microscopy, Phil mentions the challenges that all scientists face in the careful analysis of the data. Good statistical analysis is critical in biology because the measurements are inherently "noisy" due to the natural variations between the samples.

Ron and Norm applied powerful statistics because if you can describe it statistically, you can measure the uncertainty, you can actually give someone a set of quantitative parameters that if they are going to duplicate the experiment they need a match. So it was excellent science.

I wanted to go into molecular biology. I turned my attention in Norman's laboratory to the objective of mapping genes on the *E. coli* chromosome. And I ultimately did that by isolating sex factors.

As Phil's work progressed, he encountered additional technology challenges. Although not obvious to the non-scientist, the handling and manipulation of the large DNA molecules presented a challenge. However, Phil's background in physical chemistry and the handling of large molecules in solutions meant this type of work was familiar to him.

You had to understand big DNA to handle them. I understood how shear worked, how to avoid shear, I also appreciated the important chemistry when the breaking of one chemical bond in a million would fragment a large DNA molecule. So I isolated these enormous pieces of DNA [plasmids] and then used X-ray to fragment them so I could open the two strands but keep most of the strands intact.

So I developed some technology and mapped specific genes on large plasmids containing bacterial genome segments. This was novel science and I loved it, but at the end of two years I knew I didn't want to stay in the chemistry department unless it was really an avant-garde chemistry department.

Phil was sensing that more could be learned in mammalian biology now after having completed his work on bacterial DNA. Moving to mammalian cells was a big transition, but Phil thought that the technology and expertise he developed could open new understandings in mammalian systems, and he was correct.

I wanted to go into mammalian cells; mammalian systems where I thought the same technology could be used in the same way to answer other problems. There was really nothing known about genes in mammalian cells, about gene structure. I thought I could use animal viruses as biological tools with electron microscopy and other physical tools to study the structure of genes in mammalian cells.

So I moved to Cold Spring Harbor to work with adenovirus [large viruses first isolated in the human adenoids], and I ultimately got people [collaborators] who had adenovirus mutants involved so we got collaborative activities.

Mapping DNA in 1970 was not the automated process it is now, and it took a significant amount of time to complete. First is the development and use of specific restriction enzymes. These are proteins that cut DNA at specific locations. As Sharp said, he focused his attention on adenoviruses containing about 35,000 base pairs. Once the DNA is fragmented, it is separated by electrophoresis using agarose gels, which Sharp and his collaborators developed. He also introduced the use of a fluorescent dye, ethidium bromide, to mark the molecules in the gel so they could be visualized better.

By going through a sequence of restriction enzymes and then gel separation, the adenoviral DNA could be mapped. As mentioned earlier, DNA sequencing provides the letters of the DNA but mapping of the letters into genes provides the sentences, paragraphs, and the meaning of the sequences.

And all the time I was interested in questions such as how are these genes organized? How are they being transcribed? How are they being regulated? So this led me to map the RNAs, both those isolated from the nucleus and the cytoplasm of the cell.

I'm seeing big RNAs in the nucleus, but I'm seeing smaller RNAs in the cytoplasm—you know, there's something going on. And the field knew that there were big RNAs from cellular genes.

This sets the stage for the research leading to Phil's discoveries.

We did not know the relationship was between large RNA in the nucleus and the smaller RNAs in the cytoplasm. Excitingly, adenovirus showed the same relationships that cellular genes did. Then with Sue Berger, we isolated the viral cytoplasm messages and along with Claire Moore compared its structure to that of the viral genome, and that's when we saw, uh-oh, that it looks like something new.

It didn't show the structure I expected from the hybridization of the mRNA to the genome from which it was transcribed, it should have hybridized to the DNA in a colinear fashion. There was a piece that didn't anneal. Many other scientists had observed similar structures in the electron microscope but just assumed that it meant nothing. But we measured it, did a lot of statistics to show it was reproducible; was a certain length and always occurred in the same place.

This observation ultimately led Phil to the proof that some nucleotide sequences (introns) are edited out of the messenger RNA and to the discovery of RNA splicing. This breakthrough had profound effects on our understanding of genetics, the regulation of genes and production of proteins. It meant that there is a process by which a gene encoded in DNA is edited before making the protein, hence regulating the function of the gene.

To arrive at this discovery, Phil used and, in some cases, developed a number of technologies, which include:

- electron microscopy
- gel electrophoresis with agarose gel (advanced by Sharp, Joe Sambrook, and Bill Sugden)
- electrophoretic devices (designed and built under Sharp's direction)
- fluorescent labels for use in the gels using ethidium bromide (developed by Sharp)
- sample preparation for electron microscopy (advanced by Sharp).

In the original plan of this chapter, the author believed he would present a study of Phillip Sharp, the geneticist who was an early adopter of electron microscopy. However, the real story proved far more powerful than that. Sharp was a skilled physical chemist and technologist who applied these skills to achieve significant breakthroughs in biology.

This is a much richer story and example of concurrent engineering than first thought. Sharp in many ways was like Max Perutz, namely a chemist coming at molecular biology with exquisite tools he learned to use.

"That's my game," said Sharp. "I came to biology and collaborated with biologists to solve problems. That in a major sense is a good bit of convergence."

9 Protein and DNA Automated Sequencing

With numerous mysteries of protein and DNA resolved, researchers continued to seek methods to sequence proteins and then small strands of DNA. Early success led biologists to believe they could sequence the entire human genome. However, this is monumentally huge work as there are 3 billion base pairs within the human genome.

Therefore, sequencing of the genome meant developing a method of sequencing DNA that could scale to the size and complexity of the human genome. Although numerous researchers played roles large and small in the iterative path toward an automated method of sequencing DNA, at the heart of it was two-time Nobel laureate Frederick Sanger.

Sanger's first contribution was sequencing the amino acids of insulin (a protein), which he completed in 1951. For this work, he used gel electrophoresis and chromatography (see chapter 4). As we will explore in this chapter, he and others went on to develop new means of sequencing DNA, once again using gel electrophoresis. In his wonderful paper "Sequences, Sequences, and Sequences," Sanger describes the development of protein and DNA sequencers that he created during his enormously productive career [34].

Although it has been modified and improved throughout the years, the basic approach used from 1960 to today remains largely the same. It is based on finding a reagent that cleaves the sequence (strands of DNA, amino acids, or RNA) at a targeted site, such as an amino acid or specific nucleotide. The fragments are then run through gel electrophoresis to determine the size (length) of the fragment. The length is a measure of the distance between the cleavage sites and hence two locations of the target bases.

This strategy was the basis of sequencing technology until recent micromachining techniques created the possibility of a new generation of sequencing approaches that more directly measure molecules in sequence.

This chapter outlines the progression of sequencing methods as an example of technologies developed expressly for and concurrently with biological research. It concludes with a brief description of the Human Genome Project, which remains a marvelous example of the power and practicality of concurrent engineering.

Edman Protein Sequencing, 1950

The path that led to the sequencing of the human genome began with the unraveling of the structure and sequences of proteins. Although a number of discoveries related to proteins had occurred by the 1930s, there existed a lingering controversy as to their structure. On the one side were adherents of the cyclol structure formulated by mathematician and biochemical theorist Dorothy Wrinch.

Wrinch approached her work on protein structure from the point of view of a mathematician. This approach led her to develop a theory that combined concepts of mathematical symmetry with cyclol bonds, a relatively rare type of chemical bond [52]. The core of the model was the bonding at the middle of two peptides by a nitrogen–carbon bond. This creates a three-dimensional building block that could be extrapolated into creating sheet-like structures of proteins, which is something that had been observed.

Wrinch presented an elegant and complex model of the protein structure that explained many of the chemical observations, but not all.

In a 1939 paper published in the *Journal of the American Chemical Society* titled "The Structure of Proteins," noted researcher Linus Pauling wrote, "This cyclol hypothesis has been developed extensively by Wrinch, who has considered the geometry of cyclol molecules and has given discussions of the qualitative correlations of the hypothesis and the known properties of proteins . . . [However] It has been recognized by workers in the field of modern structural chemistry that the lack of conformity of the cyclol structures with the rules found to hold for simple molecules makes it very improbable that any protein molecules contain structural elements of the cyclol type" [53].

Instead, Linus Pauling and others favored a structure based on hydrogen bonding. A hydrogen atom is bonded with any electronegative atom such as nitrogen, oxygen, and so forth.

Meanwhile, at the Rockefeller Institute for Medical Research, Carl Niemann and Max Bergmann were also unraveling the structure of proteins. Through their detailed chemical analysis of various proteins, they saw a repeated structure that was inconsistent with the cyclol theory putting it into doubt. As the debate on the structure of protein continued, efforts to sequence the amino acids of proteins began. This line of research would contribute to nailing down the structure.

In 1950, about a year before Sanger sequenced insulin, Pehr Edman developed a new reagent that could be used to sequence the amino acids of proteins [54]. The process developed by Edman, known as Edman degradation, is a manual process in which the chemical phenylisothiocyanate (PITC) is used in conjunction with weak acids to cleave the protein in order—amino acid by amino acid. The cleaved segments are then run in a chromatograph to determine the identity of each cleaved amino acid, and the sequence is then assembled.

Not surprisingly, one of the early uses of the Edman process was by researchers Stanford Moore and William H. Stein (Nobel Prize laureates), also at the Rockefeller Institute for Medical Research, where Max Bergmann was and where Pehr Edman worked on his sequencing reagent (1947–1950).

In 1967, Edman and Geoffrey Begg at St. Vincent's Hospital School of Medical Research in Melbourne, Australia, built an instrument to automate the amino acid cleaving process. The design used a small spinning cup to spread the protein evenly over the side of the cup [55]. The degradation agents (PITC in this case) were added and the resulting fragments were withdrawn automatically by centrifugal force and identified. In 1969, Beckman commercialized the unit as the Beckman 890D [56] (figure 9.1).

With an automated means to perform Edman degradation made widely available, its use grew as well as the resultant advances in a wide range of biological research areas.

Moore and Stein: Automated Chromatography, 1958

Although Stanford Moore and William H. Stein are most known for their Nobel Prize–worthy discoveries related to ribonuclease in 1959, they are also known for their technology developments. In particular, they are credited with developing chromatographic methods for determining protein composition, inventing the photoelectric drop-counting fraction collector, and for building the first automated chromatographic amino acid analyzer [57].

Moore and Stein's story begins with Max Bergmann, when the two accepted positions at his laboratory at the Rockefeller Institute in 1939 and 1938, respectively. Bergmann had established the philosophy that the amino acid composition of the protein needed to be determined first and then the sequence of amino acids could be found. Among the first tasks assigned to Moore and Stein by Bergmann was establishing the chemical composition of protein [58].

Moore and Stein were aware of the chromatography techniques pioneered by Archer John Porter Martin and Richard L. M. Synge and believed they could use the technique for their work [58]. In particular, they believed they

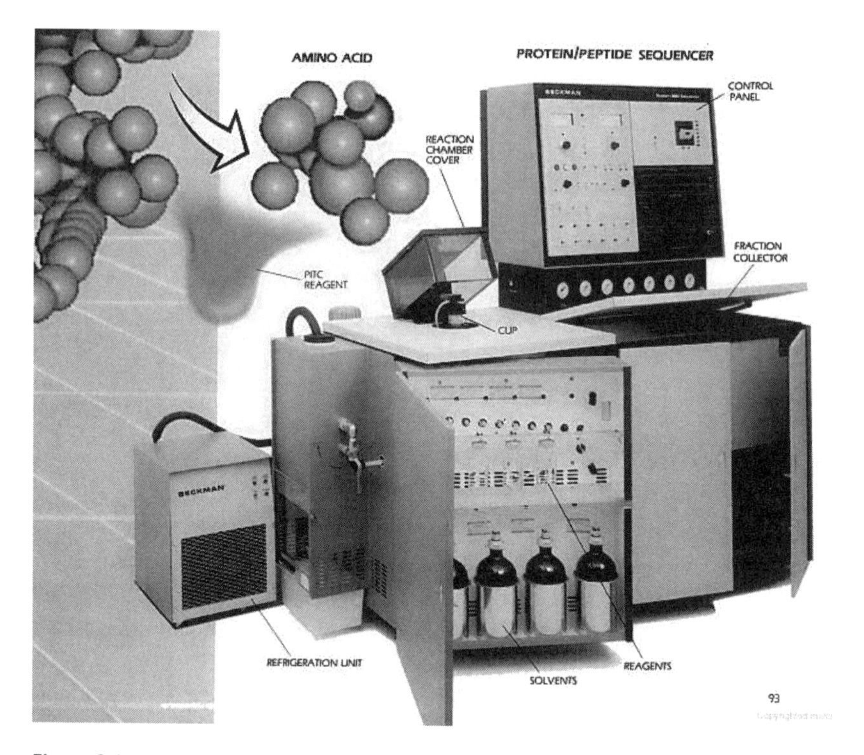

Figure 9.1
Beckman 890D Protein synthesizer offered in 1969.
Source: Popular Mechanics, March 1983, "Here Come the Assembly-Line Genes," by Dennis Eskow, p. 92.

could use chromatographic methods to determine protein composition [58]. However, they found that the (starch) chromatographic approach was slow, which led them to seek improvements. Ultimately, this led them to develop an automated amino acid chromatographic analyzer as the first step in determining the composition and then the sequence of ribonuclease (figure 9.2).

In 1956, working with 200 mg of Armour crystalline ribonuclease (made available worldwide by the Armour meat-packing company), Moore, Stein, and Werner Hirs sequenced much of the structure of ribonuclease [59]. Analysis was done in an ion exchange chromatograph. Their approach was to cleave the enzyme with chymotrypsin and trypsin and to analyze the resulting amino acids. Trypsin cleaves at lysines and arginines from the carboxyl terminal while chymotrypsin cleaves at tyrosines and phenylalanines. By breaking the peptide into fragments, then cleaving with these reagents, and finally running the sample in the ion exchange chromatograph, they determined the amino acid sequence of ribonuclease.

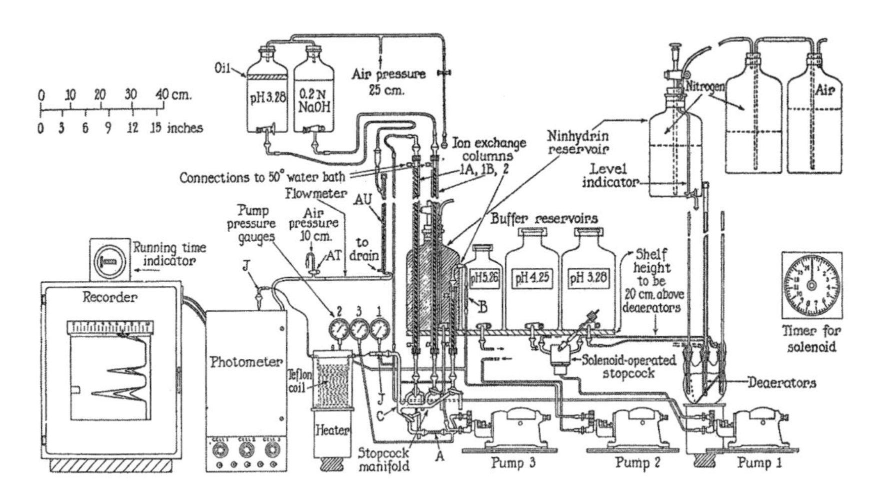

Figure 9.2
Automatic recording apparatus used in chromatographic analysis of mixtures of amino acids
Source: Analytical Chemistry 1958; 30(7):1101. Reprinted with permission from *Analytical Chemistry*.

About a decade later, Kenji Takahashi joined Stein and Moore to find the active site of ribonuclease. The team treated ribonuclease with specific reagents that eliminated the enzymatic activity, which they labeled with ^{14}C. The label was then traced by sequencing to a specific pair of amino acids.

For their work, Stein and Moore, as well as Christian B. Anfinsen, shared the 1972 Nobel Prize in Chemistry. In the autobiography Stein wrote for the Nobel Prize, he notes:

> I should like to emphasize that the development of methods grew out of a need rather than a particular desire to develop methods as ends in themselves. We needed to know the amino acid composition of proteins, we needed to be able to separate and analyze peptides in good yield, and we needed to be able to purify proteins chromatographically. Since there were no methods for doing any of these things at the time that we started, we had to devise them ourselves. [60]

DNA Sequencers, 1960–1990

By the early 1960s, the method for sequencing proteins was undergoing improvements, which included methods to automate the process better. At about the same time, a number of researchers were engaged in the proposition of determining the sequence of individual nucleotides (A, T, G, and C) within RNA and DNA. In fact, there was quite a bit of activity around this line of research.

The first substantive sequencing advancement involved RNA. This was achieved by G.W. Rushizky and C.A. Knight who published their findings in 1960 [61]. Essentially, the two men used high-voltage ionophoresis and two-dimensional paper chromatography to produce a partial sequence of the ribosomal RNA of a virus found on tobacco plants.

Frederick Sanger took the next iterative step in 1965. He used phosphorus-32 as a radioisotope label and two-dimensional paper chromatography to produce a 120-nucleotide-long sequence of samples of ribosomal RNA [62].

Then in 1972, Walter Fiers and his team of researchers completed the sequencing of a bacteriophage gene [63]. The approach they used included two-dimensional gel electrophoresis, which is slow and laborious.

Although researchers had developed methods for sequencing RNA by the early 1970s, they were inefficient and slow. A new method needed to be discovered in order to sequence DNA.

Sanger's "Plus and Minus" Method, 1975

The first big breakthrough came in 1975 when Sanger developed his "plus and minus" method for sequencing DNA. As he describes it:

> If one could produce a mixture of chains all having the same 5′ end (corresponding to the 5′ end of the primer) and finishing at the 3′ end at positions corresponding to the A residues, determination of the relative sizes of all these chains should give a measure for the relative positions of the A residues, and this, combined with similar data for the other three nucleotides, is all one needs for the complete sequence determination.

While this method of DNA sequencing was effective, it was also time consuming. Sanger could only work on relatively short sequences per batch [34]. Nonetheless, Sanger later wrote, "This new approach to DNA sequencing was I think the best idea I have ever had, being original and ultimately successful" [34].

The plus and minus method was the first of a series of DNA sequencing techniques published in the 1970s.

The Maxam and Gilbert Method, 1976

Developed by Allan Maxam and Walter Gilbert, the Maxam–Gilbert DNA sequencing method is similar to Sanger's plus and minus method. Using their new technique of base-specific chemical cleavage, in 1973 Maxam and Gilbert sequenced the lac operator, a relatively small regulatory sequence [64]. Basically, DNA was labeled at one end with a radioactive marker, and the strand was randomly cleaved with near-specific cleavage by one of four base-specific chemicals. The four chemical-treated reactions were separated by polyacrylamide electrophoresis. Unfortunately, the A reaction had some nonspecific G cleavage, so there were weak bands for sequences cleaved at G.

Following Maxam and Gilbert's work, additional improvements were made to speed up the sequencing. Sequencing of DNA or RNA was generally accomplished in a two-step process:

- The DNA was cleaved at a specific base using either a chemical (Maxam) or an enzyme (Sanger) specific to each base, followed by an addition of a radioactive or optical marker to each segment.
- The DNA segments were run in lanes of a gel electrophoresis. Each lane was for the specific base cleavage. The distance down the lane indicates the position on the DNA strand of the cleavage site and hence the base.

Figure 9.3 illustrates the Maxam–Gilbert method, though the parallels to Sanger's plus and minus method are evident.

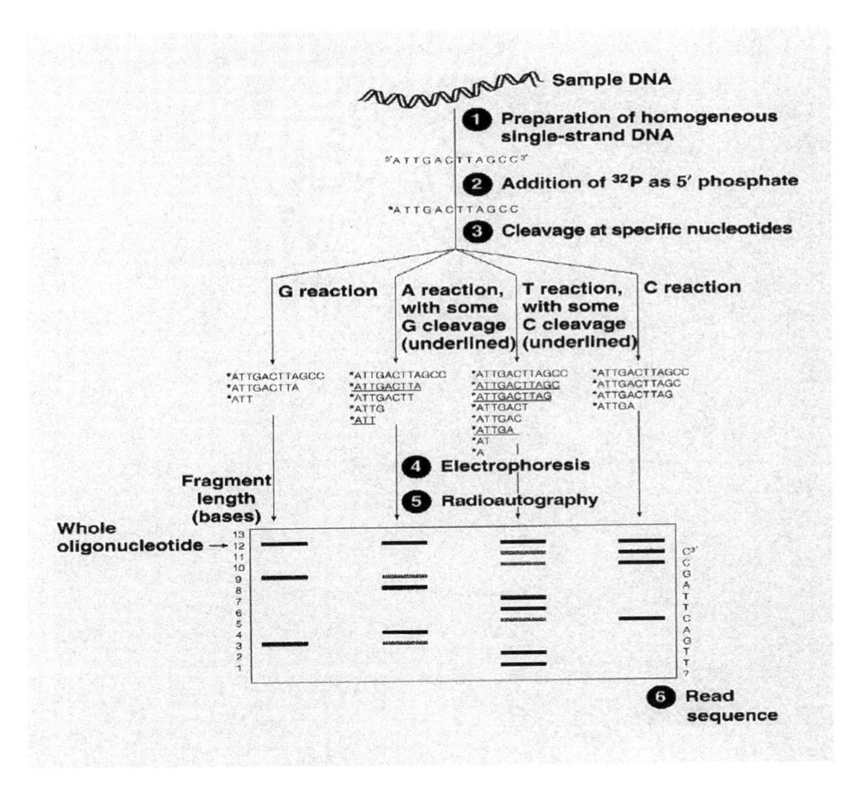

Figure 9.3
Sequencing an oligonucleotide by the Maxam–Gilbert method.

Sanger Dideoxy Sequencing, 1977

The dominance of the Maxam–Gilbert method did not last long as there were certain problems with it. Notably, the preparation was complicated to perform [65].

Sanger, for his part, was not entirely satisfied with his plus and minus method, either. As he later wrote, "While the two techniques are to some extent complementary and one must welcome any scientific progress, I cannot pretend that I was altogether overjoyed by the appearance of a competitive method. However, this did not generate any sort of a 'rat race,' and I do not think it affected our subsequent work at all. I was by no means satisfied with the plus and minus method" [34].

In 1977, Sanger developed what has come to be known as the Sanger method (also known as chain termination or dideoxy sequencing) (figure 9.4).

Sanger improved past approaches by using dideoxyribonucleotides, which, once added, would terminate the DNA synthesis reaction at a specific and known nucleotide. Each dideoxyribonucleotide (A, T, C, or G) was added in a separate tube, which contained the DNA sample, along with all the compo-

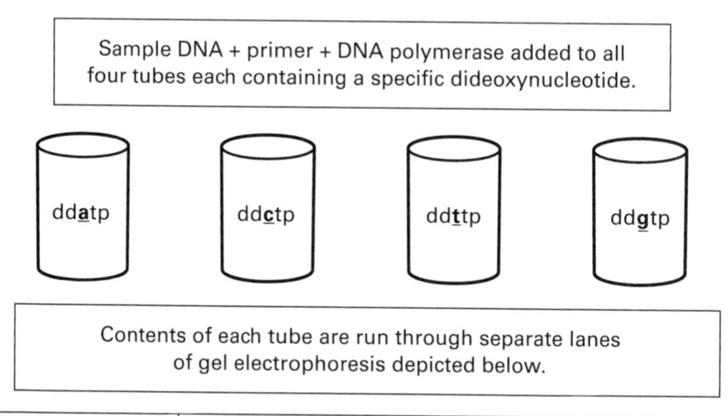

A lane	C lane	T lane	G Lane	Determined sequence
				C
				G
				G
				A
				T
				C

Figure 9.4
Example of gel result from Sanger dideoxy sequence-Sanger method. Results of the DNA replication terminated at a base (A,C,T,G) in 4 tubes are run on 4 lanes of the gel. The position along the gel represents the position (length from end to base) of the base.

nents required for DNA replication, including a primer to initiate DNA replication, the DNA polymerase, and free nucleotides. The replication of the DNA sample was initiated and would continue until a dideoxyribonucleotide molecule was placed by the polymerase into the copy strand, terminating further replication. Thus, in the tube containing dideoxycytosine, the reaction will terminate once a guanine on the parental strand was replicated. As a result, all the strands contained in this tube will end with a guanine–dideoxycytosine pair. Upon resolving the four independent reactions by gel electrophoresis in four different lanes, the relative position of each base within the sample DNA could be unequivocally determined (figure 9.4).

Chain termination proved so efficient and accurate that Sanger went on to sequence a bacteriophage in this case, using this method. And though his approach represented a manual representation of DNA sequencing, it laid the foundation for nearly all automated models to follow.

In 1980, he was awarded the second of his two Nobel Prizes in Chemistry, and then in 1983, at the age of 65, Sanger retired from research. Five years later he wrote, "The possibility seemed surprisingly attractive, especially as our work had reached a climax with the DNA sequencing method and I rather felt that to continue would be something of an anticlimax" [34].

Sequencing Gets Faster: Hood and Hunkapiller

Sanger's method of DNA sequencing proved a turning point. With his technique as the basis, numerous researchers in the late 1970s and early 1980s developed faster sequencing technologies. This in turn led biologists to suggest that the entire human genome of 20,000 to 25,000 genes and 3 billion base pairs could be sequenced.

Although Sanger's method with subsequent improvements was accurate and far more efficient than previous methods, scaling it to meet the challenge of the human genome was no small endeavor.

As Eric Lander, an MIT professor of biology and former world leader of the Human Genome Project, noted in an interview for this book, sequencing methods in the mid-1980s were relatively primitive.

> When the genome project was first proposed, around 1985, sequencing was done with slab-gel electrophoresis, which was a slab of acrylamide between two glass plates.
>
> It was done with radioactively labeled nucleotides. You stood behind a shield, you loaded stuff, and you ran it for a while. You pried off the top glass plate and you put Saran Wrap around it. You put an X-ray film on top of it. You exposed the X-ray film in the freezer. You took it off. You took your Sharpie, and you started calling (calling

out the position of the bands in the lanes). There were multiple engineering challenges along the way.

Relying on such a cumbersome process would take decades to decode the human genome. Essentially, researchers were using a slide rule when they needed a computer. Enter Leroy Hood and Michael Hunkapiller.

The 1986 Hood–Hunkapiller method used four different fluorescent dyes to label each nucleotide, allowing electrophoresis to be run in a single lane [66]. In addition to the Hood innovation, the electrophoresis could be performed in small, gel-filled capillary tubes. While a seemingly simple improvement, marking each of the four DNA bases with a different color dye vastly improved the ease and speed of sequencing. All bases could be run down a single lane, and therefore multiple parallel lanes could be used to increase the throughput. Detection with a laser/camera unit could be highly automated as the bright colors could be detected with great precision.

As with most technical breakthroughs, the development of the Hood–Hunkapiller machine got its start as researchers sought engineering solutions to support their work. Hood retells the story in a 2008 paper in the *Annual Review of Analytical Chemistry*:

> In 1982, I assembled a team including a chemist/laser expert (Lloyd Smith), an engineer/chemist (Mike Hunkapiller), a biologist with knowledge of computer science (Tim Hunkapiller), and myself, a molecular biologist—and one spring day we had a transforming conversation. Four central ideas emerged about a new proposed approach to automated DNA sequencing: (a) The DNA fragments of the Sanger reactions could probably be separated nicely by capillary gel electrophoresis. (b) The fragments could be labeled with one of four different fluorescent dyes, according to which base terminated the fragment. (c) All four colored bases could be detected together in a single capillary channel (the manual sequencing approaches used the same radioactive reporter groups that required four separate lanes, one for each DNA fragment ending in a distinct base), thus standardizing the DNA fragment comparisons and increasing the efficiency of the sequencing process. (d) The four distinct classes of fluorescence-labeled DNA fragments could be distinguished by laser detection of the dyes, and this four-parameter dye space signal could be converted into DNA sequence by computational algorithms. [67]

Leroy Hood and Michael Hunkapiller had arrived at the notion that being able to sequence DNA rapidly would accelerate all of their immunology work [68]. Their initial goal was to develop a process and instrument that separated DNA into segments to match with known sequences of complementary RNA. Additionally, they wanted it to be able to sequence a million base pairs per month.

Hood outlined his program to the dean of the School of Science at Caltech but was told the team would not get support from Caltech. The biology department viewed the focus on engineering as inappropriate for funding. Additionally, both the National Institutes of Health (NIH) and the National Science Foundation (NSF) turned down repeated proposals from Hood.

The team decided to push ahead, even though the lack of institutional support plus the technology challenge made it a daunting project.

The first major conceptual step was reached after months of brainstorming on the basic concept. At this point, the team agreed that four fluorescent dyes needed to be developed that would tag the four base molecules of the DNA. They also agreed that the basic process of detection would be fluorescent detection of the base elements.

The next milestone came 12 months later, and the first building block fell into place when Hunkapiller showed the feasibility of the four-dye nucleotide labeling.

Hood took these very promising results and returned to the NIH to appeal for funding. However, the NIH, with its historical focus on biologists in the laboratory rather than on engineering solutions for biologists, turned Hood down.

Hood and Hunkapiller decided to push ahead by looking for support in the private sector. Unfortunately, all 19 companies they approached turned them down flat because of a perceived lack of commercial appeal. Thus, the technology was viewed by traditional science (NIH, Caltech) as too much engineering and by traditional venture capitalists as too much science.

Finally, the team set up their own company—Applied Biosystems—to develop the world's first automated gene sequencing machine. Hunkapiller left the Caltech payroll and moved over full-time to Applied Biosystems, though his former laboratory and new company remained closely allied.

A prototype was completed in 1986 and was tested by Craig Venter in his NIH lab, which was the first such vetting process ever engaged in by the NIH to test a private company's product. Before long, it became clear that the ABI 377—as the machine was named—was a huge leap forward.

It is also interesting to note that in 1987, Hood and Applied Biosystems took advantage of a new National Science Foundation program called the Science and Technology Centers. The purpose of the program was to provide long-term funding for potentially transformative, complex research and education projects through partnership between academic institutions, national laboratories, industrial organizations, and other public or private entities [69]. Additionally, funding was designed to support projects where two or more disciplines (e.g., biology and engineering in this case) overlapped.

Throughout that decade, Hunkapiller continued to improve the product. His efforts eventually led to a machine that could produce a million bases of DNA sequencing in a 24-hour period, with only abound 15 minutes of operator intervention per day. If Applied Biosystems's wondrous machine—now renamed the ABI PRISM 3700, available at a cost of around $300,000—performed as anticipated, then it would take several hundred 3700s only a few years to make a first pass through the entire human genome.

Importantly, the commercial appeal of this technology was equally as impressive. Between 1987 and 1997, 6,000 of these machines were sold.

In no uncertain terms, the success of the automated DNA sequencing machine shows how important concurrent engineering is to biological research. It also demonstrates how and why funding models should adapt in order to promote concurrent engineering research models. In a speech in Kyoto, Hood incidentally makes the case:

The development of the automated DNA sequencer required a blend of technical expertise: biology, chemistry, engineering, and computer science. By the 1980s, my laboratory had developed a then unique (for biology) cross-disciplinary culture where the biologists and technologists communicated effectively with one another. In 1987, we applied for and received a newly initiated National Science Foundation (NSF) program named the Science and Technology Centers (STCs). The purpose of the these centers was three-fold: (1) to integrate science and technology (in our case molecular biotechnology), (2) to establish meaningful scientific partnerships with industry, and (3) to support educational outreach programs, which we defined as K–12 science education. The key to making this program work was 11 years of flexible funding at the level of $3 million per year (with competitive reviews). I believe the STC program was one of the most outstanding ever funded by the federal government. Our STC was exceptionally successful—we, together with others, pioneered DNA oligonucleotide array (Chip) technology and virtually started the field of proteomics. [70]

The importance of Hood and Hunkapiller's contributions to the technology, says Lander, is hard to overstate.

Really, Lee Hood played a very important role here in developing fluorescent nucleotides that you could image with a camera as they came down the lane, rather than radioactive nucleotides. When you're not radioactive right off the bat, it makes the whole process simpler. When it's fluorescent, you can image in real time, rather than leaving something against an X-ray film for a very long time, and therefore run the gel longer. You have more bases off, because it keeps running out the bottom. And just watch them as they go by.

They still have these things sandwiched between two glass plates, and you load them, but now they were nonradioactive, and they could run a long time, and you could have an automatic collector. That I think Lee Hood deserves great credit for, and the company, Applied Biosystems, that really developed the machine and had a 90-plus percent market share in the early days as a result.

But that turned out to be only a part of the Human Genome Project.

Human Genome Project

The Human Genome Project (HGP) offers a different perspective on the joining of engineering and biological research. The time frame of invention and discovery were highly compressed by closely coupling the biology and engineering in an integrated funded program. In particular, the program included the adoption of novel and risky approaches to the key parameters of work flow (sample preparation), organization, funding, and technology development.

A particularly distinctive feature of the HGP was the early recognition that very-high-speed and low-cost DNA sequencing technology would need to be developed in order to sequence the human genome. Sample analysis cost, precision, and repeatability were laid out as critical objectives for the technology to support the biology. Notably, these are requirements that were foreign to most of the genetics scientists of earlier decades and even to many of the scientists involved in the HGP when they first joined that effort. However, cost and processing speed are common criteria for engineers who embraced the challenge.

The story begins with the 1980 joint paper by David Botstein, then a genetics professor at MIT; Raymond White and Mark Skolnick, both geneticists at the University of Utah; and Ronald Davis, a Stanford biochemist and geneticist. In this paper, the authors proposed "a systematic approach to finding and organizing markers on the human chromosomes. A map consisting of such markers spaced throughout the chromosomes could then be used to locate genes" [71].

This kicked off nearly a decade of debate about the importance of, and most promising approach to, the sequencing of the human genome. During that intervening decade, several critical technologies were developed by visionaries in the field. In all, these breakthroughs made the HGP a substantially different scientific enterprise than any that had preceded it in the study of genetics.

Department of Energy Funding

As one of the earliest proponents of the HGP, the Department of Energy (DOE) recognized the challenge and sponsored the development work to reduce the unit cost of sequencing the genome. This was laid out at a 1986 workshop in Santa Fe, New Mexico, sponsored by the DOE's Office of Health and Environmental Research. At this meeting, attended by a number of key researchers, the group discussed the parameters of what would become the HGP. In particular, members discussed the technologies to be used, the expected benefits of the project, how the project would be organized and orchestrated, and how it would be funded [72].

On this latter point, attendees proposed that funding should include the DOE's Office of Health and Environmental Research as well as traditional sources such as the NSF and the NIH. Additional partners, they added, should be found among the private sector, foundations, and international organizations. Later that year, the DOE took the lead by providing $5.3 million for pilot projects conducted at the DOE national laboratories to begin developing critical resources and technologies [73].

The Santa Fe group also examined the need not just for a scalable method of sequencing—Hood and Hunkapiller were a year or so away—but also for a system that would be affordable. As the group noted in its report on the meeting, "A major emphasis concerned the rapidly changing current states of sequencing technology. While current costs may be in excess of one dollar per base pair, rapid progression toward more cost-effective sequencing (a penny or less per base pair) was envisioned simply as a consequence of the extension or combination of existing technologies and/or the development of new strategies" [72].

The group recommended placing the HGP's technology focus on those sequencing systems that met the need for very high speed, precision, repeatability, and low cost.

High-Volume Sequencing

The core challenge in decoding the human genome lay in grinding out the exact gene sequences on the genome. This had been done more or less by hand in the case of simple organisms, like bacteria, but the human genome was thousands of times more complex than that of bacteria. By the work of Maxam and Gilbert, Sanger, and then Lee Hood, the basic approach to DNA sequencing had been developed and refined.

With the fundamental sequencing technology in place, work could begin in earnest on automating the sequencing effort. Several institutions embarked on the project concurrently, including the MIT's Whitehead Institute, which started its work around 1990.

However, even before the MIT team, led by Eric Lander, could address the challenge of high-speed sequencing, its members had to think about additional engineering challenges. These included the sample-preparation process, which would require the development of new automated technology.

It is also important to note that Lander and his group were not the only researchers working on this problem. Though the HGP was international in scope and included institutions in Japan, France, Germany, and Spain, the principle work to sequence the genome was done in the United States and United Kingdom. The primary sequencing sites were Washington University in St. Louis, Baylor College of Medicine in Houston, the Whitehead Institute at MIT, and the Sanger Institute in Cambridge, England.

Among the various institutions, there was a collegial sense of competition in the work as each faced challenges related to efficient end-to-end throughput from sample preparation to analysis. As Lander notes, the most prominent challenge was automation of sample preparation. "Sample preparation was a major, major problem for all of us because we all had to maintain high enough throughput with high enough quality control to actually do the genome," said Lander in an interview for this book. This is a monumental task when you consider that the genome is 3 billion base pairs, and, as Lander noted, researchers needed to run the sequences about 10 times randomly in order for it to work. "This means that 60 million samples must be processed," he added. "So that is not well matched to, say, graduate students [to perform that work]" [74]. "So rather riskily, we put a lot of effort, over a number of years, into developing automated procedures. A lot of it was these harrowing periods of investing in technology, and knowing that if it failed, it would be a *mess*. But when it worked, we scaled up 15-fold in less than a year."

Meanwhile, the Sanger Centre chose another approach. "The Sanger Centre adopted a model of doing this all with, as they call them, *School Leavers*, which are, you know, dropouts," said Lander.

The result of these two strategies, said Lander, was a friendly contest between Lander's group at MIT and the Sanger Centre. Lander and his team insisted on dealing with the throughput issue via engineering while the Sanger Centre hired more and more people. By contrast, the Sanger Centre's method required hiring the *School Leavers*, paying them, training them, and enforcing quality control.

Lander goes on to describe their technique for sample preparation.

> Now the ways of preparing DNA involved phenol chloroform extractions; which involved mixing things with phenol and water, shake them up, and the proteins go into the phenol, and the DNA goes into the water. And then you pipette off the water phase. [Then] You've got to centrifuge it. And so centrifuging it and pipetting so you don't take the interface is, again, not highly automated stuff.
>
> So we ended up developing a bunch of bead-based preps with beads that could capture DNA. You crack cells open, put in these beads, and the beads would bind to the DNA. [When] You change salts, the beads would come off the DNA.

Moving forward, Lander and his group found a number of methods for automating their work. "We ended up with a whole lot of automated methods for doing things, with the notion that once we could get them, we could scale them," said Lander.

So what we had was the whole thing was set up with conveyor belts going back and forth. It had a relatively small staff running it, you know, eight or a dozen people [ran] this operation. It had pretty rigorous quality control throughout the whole thing. We'd monitor variability and we kept things literally in control. And we went and hired engineers, real, honest to God engineering type people. We processed a quarter million samples a day through this. This was not your typical biology.

So in some sense, the machines were an easy part, the sequence detectors were an easy part. However, setting up an operation to process a quarter million samples a day was a serious business. And then setting up an informatic system that would be able to capture and assemble this information [was another challenge]. The whole thing was run with a system so that we knew where everything was in the lab.

Lander also had to manage the funding for his project. As he notes, the Sanger Centre had a stable source through the Wellcome Trust, a British charitable research funding organization. Meanwhile, Lander was forced to compete continually with other institutions for funding.

Polymerase Chain Reaction

While not a sequencing technology, polymerase chain reaction (PCR) amplifies the DNA to be sequenced increasing the detectability. The story of the development of PCR has been widely disseminated but is repeated here for completeness.

Late on a Friday in April 1983, Kary Mullis had his ah-ha moment while driving in his car in Northern California [75]. Before long he developed PCR. Then in 1993, Mullis received the Nobel Prize in Chemistry for his breakthrough.

Mullis had been focused on developing a new process for sequencing DNA. While thinking through his concept for sequencing, he hypothesized a hybridization step in which two short primers, complementary to the 5′ and 3′ ends of the DNA template, were used along with DNA polymerase during DNA replication. Each round of this reaction would allow the DNA polymerase to copy both strands of DNA, thus in effect duplicating the amount of starting material.

"I had been spending a lot of time writing computer programs," he said, "and had become familiar with reiterative loops, procedures in which a mathematical operation is repeatedly applied to the products of earlier iterations. That experience had taught me how powerful reiterative exponential growth processes are. The DNA replication procedure I had imagined would be just such a process" [75].

In the car that night, he suddenly realized that if he could repeat this reaction 20 times, he would generate 2 to the power of 19 or more than 1 million perfect copies. This approach, however, required the ability to heat up and cool down the reaction multiple times. He quickly pulled over to jot down some notes for when he returned to the laboratory at Cetus Corporation.

When reflecting on the development of PCR in an interview for this book, Dr. Ron Davis, director of the Stanford Genome Center, noted: "The idea to amplify DNA was invented many, many times. It was an obvious use of DNA. The problem was that you had to initiate the way the [primers worked, which] are very fussy."

For his own part, Mullis had been working on oligonucleotide synthesis at Cetus and had access to technology for synthesizing short strands of DNA. This put him in a position to experiment with different starting materials for the replication process. In rethinking the problem and having the oligonucleotides already available, Mullis succeeded.

Following Mullis's ah-ha moment, the first demonstration of the PCR was November 14, 1984 [76]. A patent was filed on March 28, 1985, and Cetus joined in partnership with PerkinElmer later that year and then Kodak in 1986 for commercialization. In 1987, the first commercially available PCR product came to market [76].

The automation of the PCR process was undertaken by Cetus and its licensees. The first machine built at Cetus did not have the advantage of a temperature-resistant polymerase. Thus, new polymerase had to be added manually after each cycle to replace the polymerase denatured by the heating process. The heating process was required to dissociate the DNA strands from one another in order for the primers to anneal. Members of the development team at Cetus were John Atwood (electrical engineering), Peter Barrett (chemistry), Joseph DiCesare (biochemistry), David Jones (mechanical engineering), and Richard Leath (electrical engineering) [77]. With the discovery of temperature-stable polymerase (Taq), PerkinElmer developed the first automated PCR in 1985 [77].

In addition to Mullis winning the Nobel Prize in Chemistry in 1993, examples of the use of PCR in Nobel Prize–winning work followed quickly. For example, Oliver Smithies discovered the principles for introducing specific gene modifications in mice by the use of embryonic stem cells. In this work, published in 1988, Smithies says, "The polymerase chain reaction (PCR) described by Kary Mullis at Cold Spring Harbor in 1986 looked to be eminently suitable for this purpose and I began to work on this idea a few months after hearing Kary talk. Again, no suitable apparatus was commercially available, so Hyung-Suk Kim and I made our own PCR machine."

Smithies was using PCR within 3 years of the first demonstration and prior to commercialization.

III

CONCURRENT ENGINEERING AND BIOLOGY

10 Concurrent versus Nonconcurrent Engineering

Over the past few chapters, stories have engendered concepts of engineering that was integral with biology research and enabled discovery. These stories are examples of *concurrent engineering*. Others stories fit a more classical approach where researchers, for one reason or another, did not have access to leading-edge tools until after their development and widespread commercialization.

This chapter attempts to give an overview of the comparative record of both and concludes that biological research could be dramatically improved and the rate of breakthrough accelerated via concurrent engineering.

Examining the period that spans from Mendel's research into the inherited traits of peas to the sequencing of the human genome, one can see a subtle interplay between the kinds of tools that are available to scientists and the kinds of discoveries that those scientists are able to make. *Powerful technology precedes powerful insight.*

Picture these tools as the snowplow making movement down a snowbound highway possible. Or, more accurately, think of the glasses on the end of your nose or of your friend's nose that open up new worlds to the wearer. The tools and technologies developed via engineering—be it in the hands of a biologist, physicist, chemist, or engineer—have the same illuminating effect.

Scientific Revolutions

As the foregoing chapters have attempted to show, new technologies enable new scientific insights, which in turn provide the basis of new discoveries. Thomas Kuhn's *The Structure of Scientific Revolutions* disputes the notion that scientific knowledge is accumulated gradually and proposes that major scientific advances are leaps or paradigm shifts that are "non-cumulative developmental episodes" [78]. He provides examples (discoveries of oxygen, Uranus, X-rays) of technologies that produced a result that was an anomaly [1] of some type and stresses the role of the scientist in having the insight to recognize the inconsistency with the current paradigm. It can be a single person (Roentgen) working with a Crookes tube and a barium platinocyanide screen or a large team (Lawrence) working with a technology the size of a football field. It is the new technology that produces the new data that produce the new insight and the advancement.

Concurrent and Nonconcurrent Engineering and Science

From our viewpoint, the history of scientific discovery in biology can be divided into two categories. There are breakthroughs in which the enabling technology was developed in concert with the scientific research. Then there are those where the technology was developed in an evolutionary way, not concurrently with the science, and often for uses other than biology.

Figure 10.1 shows on the Y axis the time from proof of concept of each enabling technology to the publication of the Nobel Prize–winning discoveries addressed in this book. The case studies this book examines show the dramatic difference between the time to discovery when the technology was developed concurrently versus that when the technology was developed as an evolutionary process.

Throughout this book, we define *concurrent engineering* as the development of the enabling technology by the same team as the biology research team and

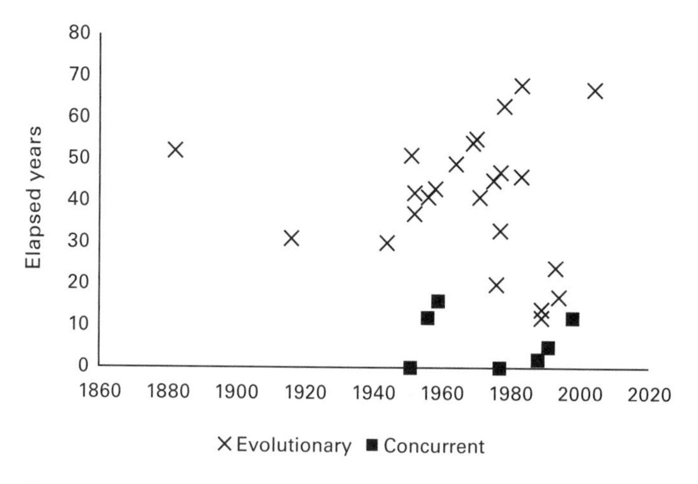

Figure 10.1
Years elapsed from the time of development of an enabling technology (proof of concept) to when it was used in breakthrough biological discovery. The figure shows the number of technologies that developed as result of natural commercial forces (evolutionary) and those developed by multidisciplinary teams concurrently working on the biology and the technology development.

evolutionary as the independent development of the technology. In the examples presented so far in this book, there were only a few that were concurrent engineering efforts, these and others will be examined in the next chapter.

Table 10.1 is the source of the data in the figure and shows the underlying data used to calculate the time span between the proof of concept for a specific technology and the year in which this technology was used to achieve a significant breakthrough in biology.

When considered in the context of figure 10.1, the remarkable observation is that evolutionary development of technology to support biology averages 40 years. Meanwhile, biological research supported by concurrent engineering can reduce that time to an average of 7 years.

This is not really a surprising result, as the evolutionary path is just that—evolutionary. In general, most of the time in the evolutionary process is spent in efforts to commercialize the technology until it makes its way into the hands of biologists. The problem here is that the marketplace creates a barrier of inertia as these technologies must be proved to be commercially viable.

Therefore, the 40-year gap in the evolutionary process is the time it takes for the technology to be seen as commercially viable and brought to the market. Whereas in the concurrent engineering scenario, the value of the technology was recognized by the biology community at the outset, and development and use proceeded immediately.

Nobel Prize Winners Are Early Technology Adopters

If one starts the clock at the moment the technology is first made generally available and clicks the timer off when the discovery is published, the time span between the two is 12 years for Nobel Prize winners and 30 years for discoveries that do not win a Nobel Prize, as shown in figure 10.2. This is strikingly less time than when measured from the proof of concept (POC), but there is an even more significant observation.

A mere 12 years from the first availability of the technology to published Noble Prize winning work is astonishing when one considers that the conduct of the research for a breakthrough discovery typically takes a decade. Then we may observe that *the Nobel laureates must be early adopters of new technology that enables their research.*

Importantly, it would appear that the start of the breakthrough research coincides with the first availability of the enabling technology. This is probably more than a coincidence, as the Nobel Prize is awarded for groundbreaking discoveries.

Table 10.1
Compilation of discoveries and enabling technologies

	Biologist	Discovery	Discovery Year	Elapsed Time, POC to Discovery (Years)[a]	Enabling Technology
1	Flemming	Chromosome	1882	52	1-µm microscope (Lister, Zeiss, and Abbe)
2	Kober	Amino acids	1916	31	Hilger spectrometer
3	Nowell and Hungerford	Chromosome translations; Philadelphia chromosome = oncogene	1960		1-µm microscope
4	Avery, MacLeod, and McCarty	Avery transforming principle	1944	30	Gel electrophoresis: Tiselius apparatus
5	Meselson and Stahl	DNA is the genetic material	1958	43	Radioisotope
6	Nathans	Restriction enzymes (not used for Sanger sequencing) used in Southern blot	1971	41	Gel electrophoresis and electron microscope
7	Bishop	Sarcoma virus protein kinase	1978	63	Western blot = gel & antibodies & radiolabeled ^{32}P
8	Collette	Phosphotyrosine kinase	1978	63	Western Blot = gel & antibodies & radiolabeled ^{33}P
9	Cavenee	Tumor suppressor gene	1983	68	Gel electrophoresis
10	Vogelstein	P53 tumor suppressor	1989	12	Gel electrophoresis: Northern blot
11	Blackburn and Greider	Telomerase RNA sequence encodes	1989	14	Southern blot (1975) and Sequenase manual sequence in gel
12	Chargaff	Chargaff base pairing	1951	51	Paper chromatography: Tsvet then Martin and Synge (Nobel Prizes in Chemistry)
13	Varmus	Retrovirus and oncogene	1976	20	Hydroxyapatite chromatography
14	Gilman	Signal conduction	1977	33	Ion exchange chromatography
15	Crick and Watson	DNA double helix	1952	42	X-ray diffraction (Bragg)
16	Marais	B-RAF structure and mechanism of activation	2004	67	Western blot, X-ray diffraction
17	Hershey and Chase	DNA is the genetic material	1952	37	Radioisotope-labeled DNA (p32) and protein (S)
18	A. Kornberg	DNA polymerase	1956	41	Radioactive tracer (de Hevesy)
19	Rodbell	Hormone signaling	1964	49	Radioisotope

Table 10.1
(Continued)

	Biologist	Discovery	Discovery Year	Elapsed Time, POC to Discovery (Years)[a]	Enabling Technology
20	Hamilton Smith	Restriction enzymes (with Daniel Nathans)	1969	54	Centrifuge (Spinco); radiolabeled ^{32}P
21	Baltimore	Retrovirus polymerase	1970	55	Radioactive tracer
22	Bishop	DNA sequence for virus			Radioisotope & cDNA
23	Stephenson	v-RAF ID as oncogene and DNA sequencing	1983	46	Radiolabeled and gel separated
24	Morrison	RAF 1 kinase phosphorylation	1993	24	Radioisotopes and Western blot, peptide sequencer Edman degradation (1969)
25	R. Kornberg	Chromatin structure	1975	45	Scanning electron microscope (Charles Oatley, Cambridge Scientific Instrument, 1965)
26	Sharp	Split gene	1977	47	Transmission electron microscope (Ernst Ruska and Max Knoll)
27	Nathans	Restriction enzymes (not used for Sanger sequencing) used in Southern blot			Gel electrophoresis and electron microscope
28	Bishop	BRCA1 tumor suppressor gene	1994	17	Single-strand conformation and sequencing
29	Zinkernagel		1985	15	Fluorescence-activated cell sorting
30	Bahraoui		1988	18	Fluorescence-activated cell sorting
31	Sanger	Sequenced insulin	1951	0	Ionophoresis prep and paper chromatography
32	Perutz	Molecular structure of hemoglobin	1959	16	X-ray diffraction, delay line storage
33	Moore and Stein	Sequenced ribonuclease (homegrown automated sequencer)	1956	12	Ion exchange chromatography
34	Moore and Stein	Gene and embryonic stem cells	1988	2	PCR
35	Smithies	RNA interference	1991	5	PCR
36	Fire	Gene silencing	1998	12	PCR
37	Fire	Bacteriophage (phi x174) sequenced	1977	0	Sanger sequencing, chain termination method (1977)

[a]The rows with blank spaces in the POC column represent breakthroughs where the researcher did not receive a Nobel Prize.

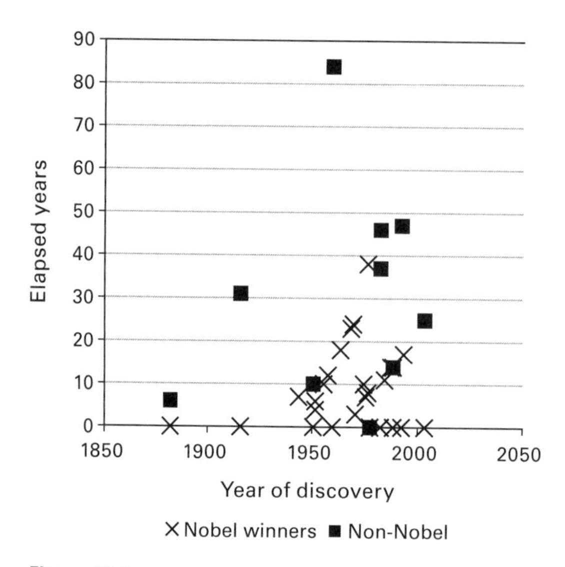

Figure 10.2
Time elapsed from the commercial availability of the enabling technology to when the biological discovery is published. Nobel Prize winners average around 12 years while non–Nobel Prize biology discoveries average around 30 years.

Multiple Steps along the 40-Year Evolutionary Pathway

The 40-year period of evolutionary progression taken by a large number of important enabling technologies used in breakthrough biological research has a number of distinctive steps along the way.

The enabling technology can be traced back to the discovery of the first phenomenon (e.g., X-rays, gel diffusion separation, etc.), which formed the foundation for the development of the technology. As such, this marks the beginning of the technology development timeline. The author observed that the evolutionary technology process develops along the following path.

1. Discovery and understanding of first physics principles that will ultimately become the basis of the technology to enable the biological research.

2. Translation of the first principles (e.g., Bragg's law of X-ray diffraction) into a POC (e.g., Bragg's father built the X-ray spectrometer for the proof of X-ray diffraction patterns), which is the first engineering step.

3. Commercial development and a series of incremental improvements to reduce cost and improve performance. Market forces pace this phase. Favorable market forces accelerate the improvements, reduce cost, and increase the distribution, which in turn helps reduce the cost still further.

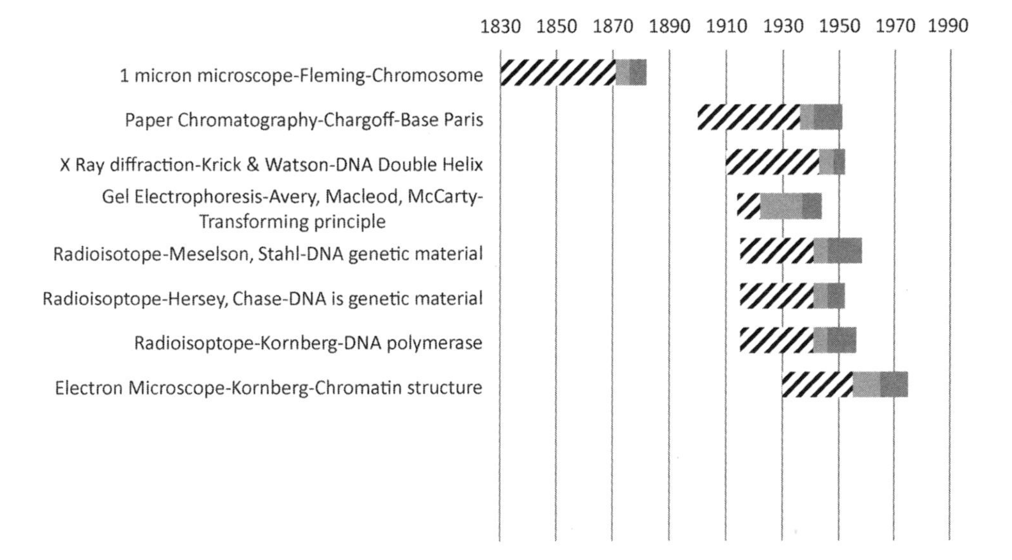

Figure 10.3
Enabling technology timeline: earliest box (leftmost) is elapsed time from technology development (proof of concept) to the commercial development, then duration of commercialization (middle box), and finally (in gray) is the time to published biological discovery.

4. The commercialization progress makes the technology widely available and financially accessible to many biology researchers who acquire the new technology in conjunction with their research, and new discoveries are made.

With these points in mind, consider figure 10.3, which shows the several stages of evolutionary development.

This figure starts with Walther Flemming at the left and then marches through two centuries of progress in the field of genetics. For our purposes, this timeline comprises seven major advances after Mendel. These milestones, which the author took from the Lander lecture on genetics, represent the pivotal discoveries in genetics. There were many more, of course, but these are the milestones the author has chosen based on the principles laid out in chapter 1.

The major biological advance is at the right-hand end of each row, shown shaded. Many of these advances resulted in glory and renown for the scientists associated with them, and ultimately in Nobel Prizes.

The leftmost block for each discovery represents the time required to develop the key enabling technology. The middle block for each represents the time it took to engineer the technology into a device that could be used by many investigators.

Again, our premise is that, for the most part, great technologies do not become available to brilliant scientists until after they are engineered into reliable devices. The purpose of figure 10.3 is to contrast the relatively short period of time taken to use the new technology for breakthrough, Nobel Prize–winning science with the long lead time to deliver that technology to the biologist.

Looking at the history of discovery in genetics and the developments of key research and clinical instruments and diagnostics, we can make the general observations that follow.

Technology Development Is Engineering

The transformation of first principles into a device that can be used by others for useful purposes (biological research or in the clinic) is engineering. Paper chromatography and gel electrophoresis are two classic examples of first principles that were transformed into technologies that could be immediately disseminated to research biologists.

By way of example, Archer John Porter Martin and Richard L.M. Synge laid out the theory and design parameters for paper chromatography in 1941. Then in a 1943 biochemical journal, Martin and Synge with A.H. Gordon presented the details of the method for paper chromatography (renamed in the article to partition chromatography). This included detailed instructions and specifications for its generalized use in studying amino acids.

Over the next 10 years, researchers in the field, including Frederick Sanger, used the technology and referenced the technique described in the paper by Martin, Synge, and Gordon. In this way, Martin and Synge not only performed the engineering to create the technology but also made paper chromatography universally available.

In a similar vein, electrophoresis (both gel and paper) was "engineered" during the POC phase, and instructions of how to duplicate the technology were published. In 1937, Tiselius showed the construction and operation of a gel electrophoresis "column," which became a mainstay in many laboratories. Then in 1951, H. Michl outlined the apparatus for high-voltage paper electrophoresis [79], which Sanger (as well as others) used in his Nobel Prize effort to isolate and sequence the amino acids of insulin.

Many Technologies Could Not Be Developed by Biologists

Many important enabling technologies could not be produced or reproduced by biologists. For example, the microscope, radioisotopes, and the electron microscope all waited 40 to 50 years before getting into the hands of a biologist in his or her research. These technologies simply could not be duplicated by the traditional biologist, and conventional laboratory equipment could not be combined to achieve equivalent performance. Biologists had to wait for the long process of commercialization to play out in order to gain access to these technologies.

X-ray diffraction is another example. Wilhelm Roentgen produced the first X-ray tube in 1901, and soon thereafter X-ray tubes were made available to the broad scientific community by the German factories of Gundelach, Muller, Pressler, and Siemens. General Electric in the United States also produced X-ray tubes at about this same time.

As a result, numerous crystalline structures were analyzed between 1913 and 1920 using available X-ray tubes and by adapting the imaging technology individually. Then in 1923, Alex Muller produced X-ray diffraction patterns of the first biological molecule [80]. This breakthrough was followed by the work of Dorothy Crowfoot Hodgkin, who determined the crystal structure of penicillin in 1949, along with several other complex molecules. She was awarded the Nobel Prize for her work.

However, while X-ray components were available as laboratory equipment prior to 1920, the author speculates that discovering the helix geometry of DNA required more precise technology. This was achieved by the Philips diffractometer and the associated X-ray diffraction algorithms developed by Arthur Lindo Patterson, which were not available until late 1940.

Breakthroughs Are Paced by Availability of High-Quality Technology

The availability of affordable high-functioning technology for diagnosis, analysis, and synthesis paces the progress of high-impact life sciences breakthroughs. The availability of these technologies prior to World War II was governed by commercial interests and paced by commercial development and distribution of the technology. Typical lag-time periods between breakthroughs were 40 years during the period 1800–1920. During and after World War II, government funding began to accelerate the progress.

Despite the accelerating effects of government funding, advancement in the biosciences still relies heavily on the commercial availability of technology. The serial nature of technology development followed by breakthrough life sciences advancement has not changed significantly since the late 1800s. The notable exceptions are instances where there has been funding specifically aimed at supporting the development of technology to support the advancement of biology.

Table 10.2 gives an overview of evolutionary technologies and some of the discoveries they enabled.

Table 10.2
Evolutionary enabling technology

	Microscope	Chromatography	(Gel) Electrophoresis	Radioactive Tracers	X-ray Diffraction	Electron Microscope	NMR
First principle discovery and POC	Abbe, 1830	Tsvet, 1900 Willstätter	Schwerin, 1915	Rutherford and de Hevesy, 1915	Roentgen, 1901 Bragg, 1910	de Broglie, 1924	Lamor, 1897 Landau and Lifshitz, 1935 Rabi
Second-stage transformed to alpha or beta	Ziess, Schott	Martin and Synge, 1941	Tiselius, 1937 Michi, 1951	Geiger and Muller, 1935	Jackson, 1895 X-ray tube Knipping and von Laue, 1912	Ruska and Knoll; von Ardenne; Oatley and Stewart, 1930	F. Bloch (physics), Varian, Proctor, Packard, 1946
Commercial	Ziess, 1876	Martin and Synge, 1941 Shandon Scientific, 1957	Numerous, 1937	ORNL, 1946	Siemens, 1913 Metalix, 1927 Philips, 1945	Cambridge Instrument Company, 1939 SEM, 1965	Varian, 1952 (NMR Specialties)
Engineers	Schott	Unknown	Unknown	NA	Unknown	Unknown	Varian
Biology discovery or clinic	Flemming (chromosome), 1882	Sanger (insulin), 1954 Chargaff rules, 1951 Varmus, 1976 Gilman, 1977	Avery and MacLeod, 1944 Nathans, 1971 Sanger (insulin), 1954 Vogelstein, 1989	A. Kornberg, 1956 Meselson and Stahl, 1958 Hershey and Chase, 1952 Rodbell, 1964 Baltimore, 1970	Muller, 1923 Hodgkin, 1937 Double helix (Franklin, Crick, Watson)	Chromatin (R. Kornberg)	Damadian, 1971 (cancer tissue)

Enabling Technologies

Of the 37 biology discoveries examined in this book, 15 depended upon gel electrophoresis or paper/column chromatography. X-ray diffraction was the runner up.

These technologies have a dominating effect on this book's timeline. The engineering required to make these tools readily usable to the biology community was done in the 1940s, and to this day researchers continue to use these basic enabling technologies, giving rise to Nobel Prize–winning research.

By contrast, sequencing and polymerase chain reaction (PCR) were developed by teams, driven by biologists, and both techniques quickly progressed from first principles to genetics laboratories. The impact of these technologies on research was observed in a mere 8 years from initial concept. This is described in greater detail in the next chapter.

It is also important to note that a tortuous 40-year path from technology discovery to use in biological research is a process that only a few technologies can survive. This raises an important question. How many potential enabling technologies did not survive the development and commercialization path for nontechnical reasons?

For example, it is entirely possible that a technology could not meet commercial expectations in a non-biology application and was abandoned, yet could have had great value to biology. In this instance, not only are biologists missing out—as well as the beneficiaries of their research—but also the commercial entity is unaware of a device's true market value.

If the marketplace creates a barrier of inertia around an unknowable number of technologies that were not shown to be commercially viable, how many opportunities have been missed?

History of Concurrent Engineering

As part of the research for this book, the author looked into identifying the institutional sources of concurrent engineering that enabled biology breakthroughs, which in turn led to Nobel Prizes. To those who follow this history, it will likely come as little surprise that there are a handful of research institutions that come up time and again.

There is, of course, the Laboratory for Molecular Biology in Cambridge, England, which counts Max Perutz, James Watson, Francis Crick, Frederick Sanger, and Sir Lawrence Bragg among its distinguished researchers. The Rockefeller Institute for Medical Research and Caltech are also very well represented on the list as numerous researchers at each have made significant contributions to our understanding of biology.

We will talk about why these institutions stand out later in this book, but we posit that they represent the early protagonists of the molecular biology revolution. They jumped on molecular biology while other institutions did not, and moreover, the University of Cambridge and Caltech, in particular, had a technology and engineering orientation.

If you want to enter into this line of research, you will out of necessity jump on new instrumentation and technology in order to succeed and remain at the leading edge. You are also very likely going to improve on the technology, as illustrated by Phillip Sharp's success in improving the contrast of electron microscopy images, described in chapter 8.

This is what you have to do to enter the game as well as maintain a position on the leading edge.

Therefore, institutional moves to get into molecular biology drove adoption and further development of technologies, which in turn helped them succeed.

Finding the Engineers

It seemed important to identify the engineers that helped develop the enabling technology in instances where there was concurrent engineering. When starting this book, the author fully expected that finding the engineer behind the development of the enabling technology might be hard.

It was nearly impossible.

With a few exceptions, the development of the enabling technology took place in a company, and there were no publications to record the development. Additionally, it is not common for researchers publicly to identify the contributions of specific engineers to their work, even as they describe the technologies and related methods they used to make their discoveries.

For example, most recently, researchers at CERN announced they had found the Higgs boson, which is a monumental discovery in the world of physics. For this reason, this discovery has been announced in many, many periodicals, even making the cover of the *Economist*. Each of these stories references the team of physicists who made the discovery and describes the 27-km-long particle accelerator that enabled the discovery.

However, there is no mention of the engineers who produced the accelerator. This may sound like a lament, but it is simply a fact that in publishing, especially in scientific papers, attention is paid exclusively to the researchers and their findings, while the creators of the enabling technology remain anonymous.

A notable exception is the lecture given by Max Perutz for his Nobel Prize, where he identified a number of people, including those that performed engineering-related tasks. Additionally, there are instances, such as our discussion of the Hubble Space Telescope earlier in this book, where an engineer plays a prominent role.

We should also point out that there are many instances in this book and beyond where a biologist, chemist, or physicist performs engineering tasks in pursuit of discovery. In this sense, concurrent engineering could almost be said to be found in the person as he or she wears the hat of a scientist for his or her discovery but wore the hat of an engineer to create the enabling technology.

Compressing Time-to-Discovery

Without concurrent engineering, there is a long gestation time needed for a given technology to be developed and deployed to the biology community.

From Walther Flemming's era to the present, the time that typically elapses between a key technology's development and its commercialization where it is made available to the life sciences community averages 40 years. Meanwhile, the time to the Nobel Prize discovery (measured from the availability of the technology to publication) is only 9 years. In other words, the enabling technology takes four times as long to develop as the life science discovery that is enabled by it.

Generally, the biology research community gains access to a given technology only after it has gone through a successful commercial development process. That process never begins unless a company detects a viable market for the product.

Today, we have the potential to change the paradigm of the past two centuries. By more effectively integrating engineering and biology and providing the funding needed to develop promising enabling technologies, we can eliminate the "commercially viable" hurdle. This one step alone will dramatically compress the time between technology development and availability to researchers.

11 The Engineers and Scientists of Concurrent Engineering

This chapter is inspired by the words of Leroy Hood when he said it is important to "always practice biology at the leading edge of biology and [invent] develop a new technology for pushing back the frontiers of biological knowledge" [70].

Clearly this point of view has been the watchword of his remarkable contributions. His sentiments also characterize the technology-enabled research of the MRC Laboratory of Molecular Research (Cambridge, England), the California Institute of Technology (Caltech), and the Rockefeller Institute for Medical Research. As we have seen in the previous chapter, these three institutions are the earliest and most noted leaders in molecular biology.

However, despite the success of Hood, his peers, and these three institutions, translation of the biology into engineering objectives is a challenging step. The complexity of biology does not lend itself to easy engineering interpretation, and, as noted in the previous chapter, there has been reluctance in some quarters of the biology community to be involved in projects that include technology development. Therefore, this chapter explores concurrent engineering in life science research with an emphasis on the few people that have made it happen and the challenges they faced.

By looking at individual cases, the author sought to find an underlying principle for success in advancing biology through the development of technology. The author chose to highlight individuals that received Nobel Prizes for their work as well as others who are equally renowned in their work in biology.

The result is that each of the people chosen for this chapter offers a different foray into biological research enabled by technologies that they themselves invented or perfected.

For example, Kary Mullis created the polymerase chain reaction (PCR), perhaps, as Ron Davis says, because of a convergence of supportive technologies, which he had also been immersed in. This is distinct from the type of approach of other Nobel Prize winners such as Frederick Sanger and Max

Perutz, who focused on technology that enabled biology throughout their careers.

Then there is Stanford Moore, William H. Stein, Leroy Hood, Ron Davis, and Steve Quake, who explored the boundaries of biology by developing new technologies that afforded them a look into new mechanisms. Meanwhile, Len Herzenberg sought technology (fluorescence-activated flow cytometry) to remove a roadblock in his biological research work. And the team of Alex Zaffaroni, J. Leighton Read, and Michael C. Pirrung intended to develop one type of new technology—a peptide synthesizer—for their drug development work but wound up developing the DNA microchip array as well.

Phillip A. Sharp is clearly known for his biological discoveries, but he was driven in part by his desire to overcome the technology boundaries that he faced, as revealed in his interview. And Robert S. Langer has applied new materials for addressing unmet needs in biology research and clinical practice throughout his career.

By interviewing these individuals for this chapter, it was the author's desire to understand how they were involved in or created concurrent engineering as a means of enabling their biological research. We also sought to learn more about the institutions they worked in and how those research centers created the environment for their work.

In all, it was our hope to bring concurrent engineering to life through the words and work of these very well-known researchers.

Frederick Sanger: Protein and DNA Sequencing, 1951

Sanger's autobiographical paper entitled "Sequences, Sequences, and Sequences" gives insights into his approach to understanding the structure and function of proteins and DNA. Much of the text is devoted to the development of reagents that would both cleave and identify by color specific amino acids.

However, Sanger does spend some time discussing the atmosphere within which he worked at the Laboratory of Molecular Biology in Cambridge, England. In particular, it was, and likely remains to this day, a place where researchers were encouraged to overcome engineering-like challenges in order to achieve biology breakthroughs.

"The Laboratory of Molecular Biology, to which I moved in 1962," he writes, "we built up ourselves and it could scarcely have failed to generate excitement and enthusiasm, but that initial impetus seemed to survive at least until I retired, and there were none of the major personal frictions that can have such an adverse effect on the research output of a laboratory" [34].

Moore and Stein, 1958

Stanford Moore and William H. Stein, along with Christian B. Anfinsen, received the Nobel Prize in Chemistry in 1972 for "their contribution to the understanding of the connection between chemical structure and catalytic activity of the active centre of the ribonuclease molecule."

This work centered on the chemistry and related biology of the structure and function of ribonuclease and deoxyribonuclease. However, in their Nobel lecture, Moore and Stein also describe the "automatic recording apparatus for the chromatographic analysis of mixtures of amino acids" and the important results coming from this technology.

Along the way to their scientific discoveries at the Rockefeller Institute, and in addition to the automated chromatographic methods, they invented the photoelectric drop-counter fraction collector and the first amino acid analyzer. They, like the other scientists in this chapter, found it necessary to invent new technology to support their work.

The implications here are twofold. The first is that they were enabled via technology to achieve significant research outcomes. The second is that they were also enabled to do what others could not simply because as inventors of the technology, they had it first.

Max Perutz—Crystallographer

The Medical Research Council's Laboratory of Molecular Biology, commonly known either as the MRC or LMB, was first led by Max Perutz (for Max Perutz's research, see chapter 6). During these early years of what is now a world-renowned research institution, the LMB became a premier example of integrating chemistry, physics, and engineering with biology in order to make remarkable advances.

Perutz set the tone and fed LMB's culture by organizing the laboratory into fully integrated technology and biology teams. These mixed teams worked collaboratively to invent the required technologies (precision X-ray diffraction components, heavy atom labeling, and an early electronic computer) to support the molecular research (structure of proteins such as hemoglobin).

This methodology is laid bare in his Nobel lecture in which he lists 20 biologists and engineers that were key investigators in his Nobel Prize–winning work.

In a 1963 paper for the journal *Science*, Perutz was equally as descriptive when explaining the funding sources for his work and how critical they were. Much of the funding came from the Medical Research Council, where

"Sir Harold Himsworth, as secretary of the Medical Research Council, made sure that we never lacked any of the funds or equipment needed for our work. Without his consistent support, research on the required scale could not have been done. Dr. G. R. Pomerat, as one of the directors of the Rockefeller Foundation, helped us to cover any additional expenditure which the Medical Research Council had difficulty in meeting. Moreover, the Rockefeller Foundation supported me in the early years; without this support the work would probably not have got beyond its initial stages" [81].

In Their Own Words: Interviews about Concurrent Engineering

A number of leading scientists and engineers were available for interview by the author about how technology enabled biological research and medical breakthroughs. The highlights of the interviews are given in the flowing sections.

Phil Sharp

In 1993, Phillip A. Sharp won the Nobel Prize in Physiology or Medicine for his discovery of RNA splicing. As outlined in chapter 8, his use of gel electrophoresis and electron microscope technology enabled this discovery.

As our interview of Sharp revealed in chapter 8, he is commonly referred to as a biologist or geneticist. However, Sharp's core training in physical chemistry underlies his research and gave him tools and insights that informed his biology.

In the excerpt of our interview with Sharp that follows, he offers his perspective not just on his own work in concurrently applying technology development to biological discovery but also on the work of others. He begins by talking about the LMB in Cambridge, England, and the transition made by its early supporters, Sir Lawrence Bragg and his father, Sir William Bragg, from physics to biology.

The creation of the LMB in Cambridge is an interesting example of leadership and the blending of physics and engineering into a transition to biology. LMB grew out of the laboratory of Bragg and Bragg, with their deep thinking and principles that advanced the use of X-ray diffraction. Since they had to build most of their instruments, the environment also provided expertise in engineering.

The leader [Sir Lawrence Bragg] decides that biological structure is the next frontier of X-ray diffraction and hires Max Perutz and John Kendrew. These are two really extraordinary people, but extraordinary people hire extraordinary people.

They almost certainly recognized [Perutz and Kendrew] as talented young people who were committed to this problem of solving the atomic structure of large proteins by

X-ray diffraction. They knew from the diffraction pattern, there was enough information in the scattered X-rays.

While resources, time, and skills were valuable assets, Perutz and Kendrew also worked within a research facility where they had access to individuals with complementary areas of expertise. Importantly, the ability to engineer new technology and devices to meet their research needs was among those assets.

As Sharp continues to tell the story, he notes that he had not given that much thought to the context and organization within which Perutz and Kendrew worked until he was provided a draft of this book.

I really hadn't thought about this until I read it in the draft of this book, but they were working in a building in which they had people who were trained to make X-ray diffraction devices.

Perutz had access to one of the first computers in England to solve the Patterson map [mathematical transformation of the diffraction pattern to yield a three-dimensional map of the crystal.

The means to solve the primary technical challenge, adds Sharp, "Was institutional convergence [of engineering, physics, and biology]."

Sharp then goes on to illustrate another important example of this type of convergence.

So think about Caltech. In 1915, [Thomas Hunt] Morgan while at Columbia was interested in a genetic model organism to study genes. He was really a descriptive developmental biologist. While examining flies (*Drosophila*), he saw a white eye and concluded that it was a mutant. Red eye is the common. Recognizing that *Drosophila* reproduces very quickly, he mated the mutant and showed that this trait segregated as a Mendelian trait [a genetic origin]. Then very talented graduate students and fellows took the genetic system and made amazing discoveries about genetics.

Morgan is then recruited to Caltech where his genetics research was located next to equally amazing chemistry and physics. I am guessing that this arrangement recruited Max Delbrück to Caltech as well as Linus. Max moved from physics to investigate the physical nature of a "gene" and led the founding of the field of molecular biology with Salvador Luria and Al Hershey. Linus made major contributions in chemistry and X-ray diffraction before discovering the alpha helix protein structure. Both Max and Linus received Nobel Prizes for their science. Caltech created an environment where physics, chemistry, mathematics, and biology as well as engineering were seamlessly combined in a small community.

As noted earlier in this book, Sharp spent time at Caltech where he learned a great deal from Norman Davidson.

Norman Davidson was, I think, one of the best examples of a chemist who made a transition to biological problems. He became a member of the National Academy of

Sciences as a chemist in his forties. At this stage, Norman studied very rapid chemical reactions using a shock tube. His teaching was on the subject of statistical mechanics and he wrote a textbook on the subject that is still used as far as I am aware. He was in every aspect a card-carrying chemist but in his late forties turned to biology.

After a sabbatical at Harvard talking to both neuroscientists and molecular biologists, he decided to take a chemical approach to studying DNA and molecular biology. This resulted in a series of papers on cooperative transitions using the principles of statistical mechanics, thermodynamics, and kinetics as applied to understanding the properties of DNA. When I read these papers as a graduate student at the U. of Illinois, they indicated a path for a trained chemist to make important contributions to molecular biology.

So I went to Davidson's lab with this evolution in my thinking. I became a molecular biologist at Caltech with Norman but primarily mapped the genomic location of genes in bacteria using the new tools of electronic microscopy of DNA. Ron Davis was completing his degree research with Norman. Ron was the first scientist to physically map the location of a mutation on a genome, in this case a deletion of genes in the bacterial phage lambda by electron microscopy. Then I went on to Cold Springs to learn animal virology both because of the opportunity to study oncogenic processes by DNA tumor viruses and cell biology. After three years, I moved to MIT to become a faculty member in the newly established Center for Cancer Research under the leadership of David Baltimore and Salvador Luria.

Sharp then goes on to talk about the current environment at MIT in the Koch Institute for Integrative Cancer Research (formerly the Center for Cancer Research):

We're creating an environment in which students and fellows in engineering and cancer biology, who have different tools and interests and do not speak the same laboratory language, can intermingle to create a credo environment. New ideas and technology will emerge from this mixing to advance both science and cancer treatment. For example, my laboratory has collaborated with those of Sangeeta Bhatia, Bob Langer, and Dan Anderson to deliver small RNAs to cells with nanoparticles.

Sharp's thoughts on the challenges ahead in biology center on the integration of engineering, biology, and systems in order to understand the complexity of biological systems and develop engineered technology to enable the unraveling of the mysteries and to develop diagnostics and drugs.

The future of cancer biology has to have the integration of microfabrication, nanotechnology, and computation. These disciplines and others are necessary to analyze complex cellular systems and then create models that predict a system's behavior upon perturbation such as treatment with drugs or irradiation.

It's such a multidimensional problem; we are going to need these technologies and disciplines. But in my opinion you just can't build a single institute that has [the capability] to solve this complex problem. You always have to have access to other scientists and engineers across a great university such as MIT and across the world. The ideal environment is one where an investigator can ask a highly knowledgeable colleague for advice.

Ron Davis

As head of the Stanford Genome Technology Center and an accomplished biologist, Ron Davis has built an organization with a large population of engineers to help create the technology needed to push the frontiers of biology. Davis was also one the original drivers of the Human Genome Project and remains at the forefront of biology.

As outlined by Phil Sharp, Davis developed one of the first mapping methods for DNA as well as DNA molecules that carry foreign DNA into a host cell for genetic manipulation. He developed the first artificially constructed chromosomes, which can be used to clone large genes.. Davis also published on genome editing, or the ability to replace any nucleotide in the yeast genome with any other nucleotide. And Davis has published on sequence variants in the genomes of humans and other species as well as identified markers for genetic and physical mapping of the human genome.

One look at the Stanford Genome Technology Center's Web site homepage says it all, "Exploiting the natural synergism between biology and technology" [82].

And the introductory text reads, "Our center develops new technologies to address important biological questions that otherwise would not be feasible. Our successes can involve improvements to existing technologies or completely new inventions, both of which aim to increase speed and accuracy while decreasing cost. In turn, once a new technology has been developed or advanced, it can often drive the perception of what is possible in the realm of experimental biology" [82].

To our way of thinking, Ron Davis has built a twenty-first century concurrent engineering organization in the image of Perutz's LMB, albeit more focused on genetics.

In an interview with Davis, the author explored Davis's view on the process of discovery and technology innovation. In general, Davis noted, "Right now, there is not much you can go with to accelerate biology. Our goal is to come up with technology that allows you to do things you couldn't do before."

As with others interviewed in this book, Davis was cautious about the balance between directive research and more opportunistic research. In general, directive research refers to projects in which the funding/sponsoring organizations articulate the research aims and solicit proposals (common in the non-biology sciences). The question posed to Davis was how discovery could be accelerated by a directed program in which a concurrent engineering capability was present.

If you look for a directive, I suspect you could do it and accelerate it with a directive. It is basically a problem then solution, and you need to articulate both of those. It is not a matter of some years of problem and you are trying to figure out a solution. That's one path. Another path you can take is here is the solution, what's the problem? And you do both of those. Basically, you are sitting there with a bunch of problems and a bunch of solutions in front of you and match them up, that's what you are trying to do.

And also there is a factor of time. It isn't a matter that you make this list and sit down one day to try to match them up, right? This is an ongoing process and you have these solutions that you don't have the right problems for. It might be ten years later that you run into a particular problem and you say, "Ah! We figured out the solution to it ten years ago."

You keep them. You make sure you integrate that across time and the larger your list of solutions and problems, the more likely you get hits. And the longer time frame increases your chances. This is very common for us to come up to a problem we are going to take and we have the solution.

To illustrate his point above, Ron chose to talk about the development of PCR. The problem to be solved was how to amplify DNA. A single strand of DNA is difficult to analyze with conventional technology and is nearly impossible to use in genetic growth experiments. A method for faithfully reproducing copies of the DNA double strand was clearly needed to enable research of the DNA and its use in biology experiments.

The idea to amplify DNA was invented many, many times. It was an obvious use of DNA polymerase to reproduce DNA. The problem was that the DNA polymerase is very fussy. It requires a primer. [A primer is a small strand of DNA that the replicating machine of DNA polymerase recognizes as the starting point for replication.] They don't just start synthesis. How do you get primers? Oligonucleotide synthesis. [Another technology was needed, which could build primers to order from basic inorganic chemical building blocks in a fairly automated fashion.] So PCR is an interesting one because it's obvious that the problem is getting DNA amplified. The solution is to use polymerase, but there was a missing component: primers.

Oligonucleotide synthesis was becoming available by Har Gobind Khorana. [Khorana published the first synthesis of DNA in 1959 [83].] Fairly soon it became better and better. It was one of these things that you didn't go back and re-make that connection of here's the problem so where's the solution? It's like, "Aha! Now we have oligonucleotides." It required somebody else to re-think from it and then put it together. So that was an interesting one because it is an idea that many, many people had, but couldn't do it.

It [DNA replication] just hits you perfectly and then this technology of oligonucleotide synthesis became available [within] a year or two. That impacts it in many of these pictures where people had the idea [of DNA replication] and tried to work on it, but the [primer] technology wasn't there. As soon as the [primer] technology comes in, the next person who thinks of it [DNA replication] is the one who is going to get the next Nobel Prize.

So the biologists would be the ones who say this is a real important problem but have no technological skill for doing the chemistry. The chemists would say why would you want to do that? So biology had to [find the] chemistry. Khorana was one of the chemists who was interested in biology.

Davis then went on to explain one method of organizing concurrent engineering at an institutional level.

Pair up a good biology post doc with an engineering student. Do not pair faculty. The biologist gives you the problem and the engineer gives you the solution. You put the two brains together on what you are trying to do. They teach each other.

To solve the big problems, you try and get a student in an area who knows something about the big problem. I don't know if this is work for a medical student, they don't have a lot of freedom. [However] whenever I have done that, it works spectacularly well.

After bringing [the team] together, I start asking them questions to get them to think. I don't tell them about the direction, I ask them questions. That's my style. Get them to think and realize that they have the power within them to do it.

Eric Lander—Genome Sequencing

A major part of the challenge faced at MIT's Whitehead Institute for Biomedical Research—and presumably at the friendly competitor institutions involved in the sequencing effort elsewhere—was the need to sustain a constant learning process across the traditional discipline boundaries between biology and engineering.

In this effort, Eric Lander, a member of the Whitehead Institute and a biology professor at MIT, played a leading role. As his biography on the Whitehead Institute's Web site reads, "Lander was a world leader of the International Human Genome Project. . . . Under his leadership, the Whitehead/MIT Center for Genome Research (which formed the core of the Broad Institute) was responsible for developing many of the key tools of modern mammalian genomics and was the leading contributor to the Human Genome Project."

Therefore, Lander has an interesting and compelling perspective on the role of concurrent engineering. Going back to our interview of Eric:

Biologists simply don't think in terms of process engineering. I have to confess that I don't normally think this way. But we had to have a culture here where we had engineers and biologists talking to each other, and beginning to understand this business about having the little control lines, and wanting things to stay within the control lines.

Inventing a new scientific culture—roughly consisting of equal parts biology, process engineering, and computation sciences—was a huge challenge in and of itself.

At a critical juncture where Lander and his team needed funding as well as to focus their effort on developing the genome technology, the project moved into a new phase. This phase occurred with the establishment of the Broad Institute to house the resources necessary to move the sequencing into high gear.

Lander describes the effort to scale up the work at the Broad Institute:

Even when your good friends are there, you get frustrated when they tell you one thing and you don't really believe it. So we were frustrated with this idea, well, you couldn't just scale up, it just wasn't possible. And I just didn't see any law of physics that would be violated by scaling (sequencing) fourfold, or tenfold, or something. It (the notion that you could not scale up) was a law of organization.

In order to overcome the challenges of scaling their efforts, Lander recalls, "We had to organize in radically new ways at the Broad. We had a concept called platforms. Of course, we had academic labs, which come together in things called programs, but we also had platforms.

For example, we had a platform of 200 people in sequencing, none of whom were on the faculty track. We had a platform of 50 people in genetic analysis, none of whom were faculty track. We had a platform in proteomics. We had a platform in biological samples handling. All of these were professionally staffed, permanent positions, not transient positions like the techs in many labs. [And the platforms] partnered with the programs.

And what they partnered on, said Lander, was *technology*.

The platforms listened to the programs and provided the necessary technology. In some cases, that technology was purchased off the shelf and needed minor or major modification. In other cases, the missing technologies had to be developed in-house.

The concept was not exactly new. It was similar in some of its fundamentals to the traditional "core facility" methodology used broadly in the National Institutes of Health (NIH) funding system. However, it was a departure in terms of both its scale and its scope. Whereas a traditional core facility tends to operate with existing commercial equipment, the Lander platforms provided advanced development competencies capable of *conceiving*, *developing*, and *refining* the technology needed to push the biology forward. Lander had implemented concurrent engineering and biology research.

These conceptual and structural innovations—including the integration of multiple scientific languages and the adoption of unorthodox organizational designs—enabled the teams to make progress within a few months that otherwise would have taken years.

Other factors also contributed. For example, the project leaders found a way to overcome a challenge inherent in mobilizing such a large work force in academia, which is the issue of scientific recognition through publications. The traditional goal of the faculty reward system is to achieve principal investigator (PI) status, and—by extension—ownership of the scientific subject matter. By its nature, the work on the genome would have had to be a huge team effort, with widely distributed recognition. So why would tenure-track faculty risk committing their time and energy to the project?

"Enlightened self-interest," responds Eric Lander. "People were willing to sign up for teams if by being a part of the team and devoting half or even two-thirds of their effort to that team, they themselves can get five times as much done in the remaining portion of their time. That's always been a great trade-off. They get more publications and they get higher-profile publications."

With many of the larger organizational and technical problems resolved, Lander and his team continued to identify and overcome emergent challenges in their work. Asked about some of these, he said, "I'll give you the famous example, slightly after, a year or two after the Genome Project, but it was very much in the spirit of it, and our favorite example."

We found that we knew our allowable variation well enough that when we got a batch of enzyme from our supplier that was really right by the edge of what was acceptable to pass on, we decided to quarantine the lot and investigate what went on. We could tell that something was funny about this batch.

So we were able to determine that it looked like it had some pre-amplification of DNA that had gone on in it. This all had come off our automated pipelines where we're up here and this [batch] was just down here by maybe 10 or 15 percent, which nobody would have bothered to worry about. But that actually mattered to us, in terms of what our yields would be.

We sent our engineering team over to two of the supplier plants, and eventually tracked it down to a plant in Ireland, where they mixed things in a cold bucket. It turned out there was no SOP [standard operating procedure] for how long that bucket had been chilled in advance. Sometimes the bucket wasn't cold enough and the reaction took place.

Because of that the company here went and redid the whole SOP, credited us for $1.75 million worth of goods, and redeveloped the product SOP for this.

Lander discusses another engineering challenge they faced.

What we did was we actually did a technology project on the physical mapping that managed to save us enough money that we had some excess money that we used to start investing in technology for sequencing. Very briefly, we had to do millions and millions and millions of PCR reactions. So we ended up developing a Brobdingnagian system.

We had a 1,536-headed pipettor that pipetted things into a plastic microtiter plate that was sealed with a little plastic film on the bottom, and heat-sealed with film on the top. [The microtiter plates] then went into this massive casket that rode over three water baths, the water was forced through it, so we could parallel process hundreds of thousands of PCR reactions.

Then we had to get the PCR reactions out of these plastic plates. So they went on to another machine that looked like a medieval torture device. It had a bed of 1,536 needles that pierced the [microtiter plate] and a vacuum sucked [the sample] down onto a membrane and the membrane was hybridized.

So we had this [Brobdingnagian system] built. We went to a little engineering firm here and built this whole thing. At the end of the day it was 15 months to build the one thing, and then it took us 11 months to completely map the human genome by creating a physical map of the human genome.

And it saved us enough money that we were able to start investing in some of the [other] new technologies.

But of course, as Lander is quick to add, the formula only works if the organization's leaders succeed in bringing together a team that's sufficiently productive, adequately resourced, and working on topics that are big enough to spark and support multiple individual forays.

Steve Quake

While starting on this book, the author would regularly hear "You must see Steve Quake." Steve does concurrent engineering and biology research at the leading edge. He is currently a professor of bioengineering and also of applied physics at Stanford University where he did his undergraduate work. After a Ph.D. from the University of Oxford, he went to Caltech in 1996 and then returned to Stanford in 2004. Steve has developed a technology that is often referred to as Integrated Circuit of Biology. He developed chip-level systems for dynamically controlling fluids and cells to explore the behavior and structure of cells and proteins. For instance, he developed a chip that controls prefect crystallization of proteins for analysis by X-ray diffraction, performed high-throughput analysis of the interactions between proteins, which helped lead to drug discovery, designed a digital PCR device used to diagnose genetic disorders from small samples of blood, and so forth.

In reflecting on the challenge of conducting complex biology projects in the academic setting, Steve Quake talked about managing and conducting large, complex biological research: "We're not good managers, in most senses of the word, and we're not even good at working together because everyone just cares about what the next paper is going to be and who is the last author going to be and so forth. My strong opinion is that there is just some things that are better done in the context of a small company or some other organizations where the mission is clear." Asked to share some thoughts about Big Science and tackling seemingly intractable biology problems, Steve offered one large challenge: "It centers on systems biology and things like that, and centers—the Alliance for Cellular Signaling, proposed by Alfred Gilman, a Nobel Laureate at UT Southwestern in Texas. He organized a big multi-institution center, the Alliance for Cellular Signaling. And then they basically went after that problem" And what happened? "Not a lot from what I can tell."

The Alliance for Cellular Signaling (ACS) was started in 2000, largely funded by an NIH Glue Grant. The Glue Grants were started by NIH to encourage multidisciplinary teams to address large problems. Five major awards constituted the bulk of the program, and after a panel assessment the entire program was terminated in 2011. The report concluded that "the scope and impact of the scientific knowledge outcomes of the Glue Grant Awards Program as a whole were not commensurate with the investment" [84]. The project funding ended in 2007 after $63.7 million in funding. The project, which set out to map cellular signaling pathways in heart muscle cells and immune cells, was deemed to have "significant flaws." It was also the only one out of five projects that the NIH phased out early, after only seven rather than 10 years [85].

Of the five Glue Projects, "One substantial success was the Consortium for Functional Glycomics (CFG), which developed arrays, mouse models, and other resources for researchers studying sugar molecules that cells use to communicate." The stated strategy of the successful Consortium for Functional Glycomics was to "work with the scientific community to create unique resources and services that Participating Investigators can utilize." The leadership of the CFG was largely focused on bioengineering and biochemistry, and their aims were clearly to develop enabling technologies. The ACS leadership and aims, by contrast, were more centered on the biology research.

Clearly, the goal of understanding cell signaling is meritorious. The problem appeared to be in the management and execution of the effort, which, as we will discuss in chapter 14, would need concurrent engineering and biology to overcome the technology limitation in the current cell signaling studies.

Continuing the discussion about teams coming together to develop enabling technology:

You could talk about microfluidic protein crystallization. It turns out getting those crystals is the hardest part about getting the structures. There is no rhyme or reason how they crystalize. You learn about one and it gives you a very full intuition about how the next one is going to work. I had funding as a promising young investigator and every year the Fellows together for a small meeting, 30 or 40 of them in one place. The first year and the last year fellows give talks, and I was in the first year, and I gave a talk about all this cool (microfluidic) plumbing. And one of the other first-year fellows in the audience was a structural biologist who said hey, maybe this would be awesome for solving some of these problems with protein crystallization. We struck up a collaboration. First we covered feasibility, proof of principle, those type things, figured out how to get crystals to grow in small volumes, and learned how to crystalize stuff. We finally had it all going, we in one week did more experiments than the best post-doc did in a year. It was a real proprietary breakthrough. They took it right out of my lab. We built the engineering and development, and they commercialized it right away.

Asked about examples of major breakthroughs in biology and the problem of finding them:

I think there's a lot of oral history to science, to be honest. I don't know what the perfect—you know, the one that I like to use, the simple one, is you've got Crick and Watson working away, but the—it was the X-ray diffraction that gave them the opportunity—*But the thing that's classic about that is the X-ray diffractions took 40 years to get to the point where anybody could be using it.*

Well, but it took a long time before it even got to that point. In other words, you know, it was really turn of the century when Bragg demonstrated X-ray diffraction. Then the technology [instrument and computational analysis] had to be developed. That aspect of it. It's more just the sophistication of solving these really complex scattering patterns. It was driven more by the mathematics and the question of, you know, could you even crystalize a protein.

The author asked about how to fund the technology development effectively.

I actually do like the DARPA model.

Everybody who is, you know, a scientist of a certain quality gets some baseline funding, okay, enough to have allowed a couple of—a couple graduate students, let's say. And they don't have to write proposals as long as they have shown some minimal productivity or confidence. They're just funded at that baseline level, and I'll argue a little bit for this approach. And then people wanted to take on bigger things, or when things become timely, that's where you have competitive proposal processes. It's like DARPA things which are formed around exploring an idea and hitting it hard and seeing where five years is going to get you. These (DARPA) program managers, they rotate in, they're casing around for good ideas, and they call meetings and they bring together the academic community and people brainstorm, and ideas come out of it—it's a bottom up process. The program manager doesn't dictate.

The DARPA program manager does decide on the priority, and he sells it up the chain of command, because it's got to go all the way to the top and back down. At some point there is somebody that has chosen a stretch goal (frequently called "DARPA HARD"). Steve Quake again:

> The NIH is very much going in that direction now. I mean, they are having very targeted RFA's. They are telling you what you're supposed to write for, and the R01, the old school, we would send in the R01 for, you know, could be a much smaller fraction of whole.

So there are bigger procurements? "Oh, yeah, the centers and stuff like that and the RFA's, and those are becoming a much larger fraction of the whole."

While NIH is moving toward larger collaborative projects in 2011, the vast majority of the $3.4 billion in new grant awards went to traditional smaller R01 grants accounting for $3.1 billion, or 94% of the budget. With the average funding of a new R01 at $436,053, there is hardly enough funding to support two PIs, and it is very unlikely to support concurrent engineering and science. One hundred forty-three new Center grants in 2011 averaged $1,208,453 and totaled $172 million of the $3.5 billion, or 5% of the new grants. While the Center grants offer more funding opportunities for a current engineering and biological research effort, they still fall far below the kind of funding needed for the development of the technologies like sequencers and microarrays, which were funded at the $30 million to $40 million level (discussed later in this chapter).

Robert S. Langer—Biomedical Engineering

A book about engineering-enabled medical breakthroughs would be incomplete without an examination of those that cross into the clinic. There are ongoing and tremendous contributions from a large community of biomedical engineers that are affecting both biology research and clinical practices.

Chapter 14 addresses the impact of concurrent engineering contributions to the clinic. However, in this chapter, the author thought it important to look at the individuals that exemplify concurrent engineering in biological research.

In particular, there is MIT professor Robert S. Langer, who heads one of the world's largest biomedical engineering labs in the world. He has written more than 1,175 articles and has approximately 800 issued and pending patents, which have been licensed or sublicensed to more than 250 pharmaceutical, chemical, biotechnology, and medical device companies. As such, he is one of the most cited engineers in history [86].

With such a notable engineer as its leader, it is little wonder that the Langer Lab practices concurrent biology and engineering. As the lab's Web site notes on its homepage, "Our work is at the interface of biotechnology and materials science."

In fact, the lab's work in tissue engineering and drug delivery are excellent examples of pushing the scientific boundaries by concurrently developing the engineering to drive discovery. The end goal for the latter is to develop an engineered system for delivering drugs more effectively to their targets while the former seeks to develop synthetic means for generating tissue. However, both efforts involve biology research to uncover the governing mechanisms that control or limit the delivery of molecules to target cells or to foster the effective growth of tissue.

From here, we will let Bob tell his own story:

My dad and my guidance counselor told me that I should be an engineer because I was good in math and science. And so that's why I went into engineering. Maybe if I'd understood things better I would have gone into medicine, but then I don't know that I would have come up with the different ideas because really for me what made so much of a difference in my life was being an engineer and then in 1974 going to the surgery lab. It was here that I was exposed to Judah Folkman [founder of the field of angiogenesis research (angiogenesis refers to new growth of blood vessels that is highjacked by cancer to bring nutrients into the tumor to support growth)] who was a visionary . . . and I think engineers try to solve problems and so what I would see with angiogenesis was: How could we solve that problem?

Asked about his role working in the Folkman medical laboratory, Langer said, "You know, it's just like I'd sort of see these medical problems involving materials [and] my thought was, how do materials get into medicine?

If you look at pretty much every material that's in medicine in the twentieth century, its driven by an M.D., and what they did is they went to their house and found some object that resembled [what] they wanted to fix. The breast implant they picked a mattress stuffing because it was squishy. I could give you five other examples, but I wasn't an M.D., I was an engineer and I didn't think like that.

I thought [about] chemical engineering design because that's what I'd learned. I said, "Why do you go to your house and pick something? Why don't you ask the question: What do you really want in a biomaterial from an engineering standpoint, biology standpoint, chemistry standpoint, and make it, synthesize it?" So it's just a different frame of mind.

If I hadn't known chemical engineering, I wouldn't have thought about [those questions] and that drives so much of what we've done over the years.

It's Neither Engineering nor Medicine: Orthogonal Thinking

After successful postdoctoral years in Folkman's lab, Langer decided in 1977 to explore a faculty position at a major research institution. While clearly at the cutting edge of dramatic developments in bioengineering, his résumé was not conventional and he found it difficult to land a position. Unconventional is more the rule of the innovators we find in this book. It is those with unconventional perspectives that are most able to bring orthogonal thinking to the problem.

I applied to different schools, but no chemical engineering department would take me. They said, "This isn't engineering." Judah Folkman knew Nevin Scrimshaw, who was head of the nutrition department [at MIT] and he talked to him and Scrimshaw decided he would hire me. So then I went to the Nutrition and Food Science Department, but the year after I got the position Scrimshaw decided to leave. Several very influential people in my department decided to give me advice and said I should start looking for another job because they said what I was doing didn't fit with what they were doing in [our departments]. That was pretty discouraging.

During my first couple of years, I was getting nearly all my grants turned down. Even though we'd published some of the stuff, I don't think a lot of people believed that this (materials science applied to medicine) would really change the way you would use materials. What happened is I just kind of hung in there and eventually I did get some grants. Some of it is just putting your head down, and eventually people started using what I did. [It was the] pharmaceutical industry that started using what I did and they spoke out and they actually would tell people, "Hey, you know, this is really good stuff."

Concurrent Engineering and Discovery: Tissue Engineering

Asked about the first efforts in engineering of human tissue and organs, Bob explains:

Well, the general idea goes back to close to 30 years ago. I have a good friend Jay Vacanti, now he's head of pediatric surgery at Mass General Hospital. So he came to see me, it must have been 1983 when he was head of the liver transplant program at Children's Hospital, and he was seeing kids that were dying of liver failure. He said, "Could we come up with a strategy where we could make tissues and organs from scratch?"

I started going back and forth with him on this idea that there are polymer scaffolds that you can put the cells on. At that point, people weren't talking about stem cells or anything like that and so we used normal differentiated cells.

However, as their work continued, it became apparent that Langer and his team weren't able to get enough surface area to work with by using two-dimensional scaffolding.

So Jay, I remember one day, he was on a beach and he calls me up. He says, "Can I have the polymer scaffold [something] like seaweed?" And I said, "Sure we can make that." So we did.

So we designed a three-dimensional scaffold sort of like seaweed [with] all the right shapes.

Before long, Langer published his results in a number of papers, and eventually this helped lead to what is now known as artificial skin, which has been approved by the Food and Drug Administration (FDA).

Engineering Blood Vessels

Langer's attention next turned to making artificial blood vessels.

Well, we were trying to make a blood vessel, and this was a paper we published in *Science* 10 or 12 years ago, and so we were trying to make a blood vessel and it wasn't working. People had tried to make blood vessels for years, [but] nobody had ever succeeded.

When they (biologists) grow cells in culture, the cells just sit there and hang out in a Petrie dish or a flask. Maybe there's a little motion, but they just sit there. And my postdoctoral fellow Laura [Laura Niklason, M.D., Ph.D.], when we started talking about [the problem], you know, she thought, "Well, but that's not what happens in the body. The body would have these blood vessels exposed to a pump like the heart." And she said, "Well, maybe we can make a bioreactor." That's what people in our lab were working on. And then she said, well . . . so let's make some bioreactors where these little blood vessels we were trying to make [were] hooked up to a pump like your heart rather than just have them sit in a Petrie dish.

You basically pulse the media through these little blood vessels beating like a heart and that totally changed everything. So then we were able to make a good blood vessel and we published that in *Science*. But the bioreactors are really interesting [because] that *Science* paper in 1999 was just a simple example of using engineering. It was a different way of thinking because everybody had used tissue culture regularly, [but not with a] bioreactor.

Bob has continued to apply engineering principles to difficult biological problems as evidenced by his recent work in developing nanoparticle-based drugs for targeted therapy. Bob is always right at the interface of addressing big biology challenges with leading-edge technology and engineering.

Len Herzenberg—1970 Fluorescence-Activated Cell Sorter

We will pick up the story of the fluorescence-activated cell sorter (FACS) in 1959. This is when Len Herzenberg completed his graduate work at Caltech and moved with Lee, his wife and scientific collaborator, to Stanford to work with Joshua Lederberg. At this point, Lederberg had won the Nobel Prize in Physiology or Medicine only the year before.

The area of study was immunology in mammalian organisms. Len and Lee attached a fluorescent molecule to antibodies designed to recognize specific molecules that stick from the cell surface and can be used to distinguish one cell type from another. They then could observe and count different cells manually under a microscope. However, this manual process proved slow. Len's intuition led him to look for a technology that could sort the cells at a higher speed.

I went up to Biochemistry [at Stanford] and talked to David Hogness, Dale Kaiser, and Roger Kornberg, and said "Isn't there some kind of machine, which could look at these fluorescent cells and sort them or count them and then sort them?" And they said the closest they even thought of, all three of them, was at Los Alamos where they were studying particles that are generated from atomic bomb testing in the atmosphere and go up into the sky, into the mushroom cloud. [They] send the animals up into the atmosphere that breathe in some of the dust and the dirt that's become radioactive. Then in Los Alamos they would sort these particles by volume and based it on their volume.

I went to visit them in Los Alamos and I said, "You know, you guys are good engineers, why don't you just add fluorescence to the means of detecting the objects in cells or particles? And then that would be fantastic for immunology and cell biology." They said, "Look, it's not in our mission. Our mission is to study fallout and atomic bomb testing." And so after spending a day and a half there, they were not interested.

Herzenberg returned to Stanford empty handed except for an idea that a high-speed cell sorter could be done. Recall that the only instrument for cell analysis at the time was the Coulter counter, which only counted cells based on size characteristics.

I was here at Stanford and Lederberg had initiated the Mars/XOBiology project to look for life on Mars and so I was playing ping-pong with these engineers. Literally ping-pong because they were located right opposite me in the basement; I would have lunch with them and play ping-pong.

I walked into the dining room and I said okay, I went to Los Alamos and after they said they are not going to work on fluorescence I said, well can I take the plans back to Stanford and see if I can't get an engineer or two to help me. So I got these guys I was playing ping-pong with. And [Los Alamos] said yes, I can have the plans and I'll try and reproduce what you have with the volume separation/volume projection and then I'll try and add fluorescence to it. The guy who helped me to add fluorescence was studying the amount of fluorescence that got bound to RNA or DNA.

You know what the inkjet printer is I'm sure. That's the principle upon which this is [based]; the droplets are separated and the principle of making the droplets for the inkjet printer came from Lord Rayleigh in 1844. [He] found that if you shake a nozzle of a hose you are going to get droplets coming out; use a piece of electric crystal to do it [and] you are going to make 30,000 shakes a second. And that's how we can make 30,000 droplets a second [with] the right diameter. So I got the guy who invented the printer to come work with me.

Herzenberg imagined that if cells were in solution then they would be captured in the individual droplets, which could then flow past a detector. The cells of interest would be labeled with a fluorescent marker and the detector would signal an electrostatic section to "pull" that droplet containing the target cell out of the flow.

When asked about his motivation to develop FACS, which may seem pretty far afield from his original focus on the biology of the immune system, Herzenberg says his motive was "To deal with the problem; the scientific problem."

And Lee Herzenberg says about Len, "He does not make a distinction between the two [the biology and the engineering]. He absolutely does not. It is not in his vocabulary to make a distinction between them.

I didn't do any of the engineering on this project. However, I was deeply involved in the daily development. I was essentially the head of the design team and took responsibility for assuring that the machine would be useable by scientists doing immunological or genetic studies.

Once the prototype was working, Herzenberg wanted to achieve higher-speed sorting.

So when they finally got this machine going, I said I wanted it faster. They said "Well, we will have to repeal the laws of physics." And Len said, "I don't care about the laws of physics, but if you want to do biology . . . you've got to get it [sped] up."

And they figured out a way to repeal the laws of physics. This was his driving force, it had to be able to get biological data to answer real questions.

Once the speeds of interest were achieved, Herzenberg looked for a manufacturer.

I found this company, Becton Dickinson, and there was a guy out here in Atherton, nearby, who came by and said could I consult with them on some immunology project. And I said "you know the thing that we're developing is the most important thing to immunology that's coming along, the FACS." After I convinced him that this was the case, he said "Well, I've got to get it through the board of directors at Becton Dickinson, which is in New Jersey."

The first device was completed around 1970 [87], and Becton Dickinson (BD) built and sold the first units around 1976 [88]. "BD would have walked away from the venture if I hadn't gotten an NIH contract that would let me subcontract the building of two such instruments to Bernie's group [BD] and let me collaborate in the effort [1974]," added Herzenberg.

He goes on to note, "You know what we're taught here in this lab as an active member? Don't tell me the technology you want; tell me what you are trying to do. And that's our engineers who would always look at this and say never mind telling me how to do this [because] we are limited, biologists are incredibly limited by what they think you [engineers] can do."

The Herzenberg group went on to do two decades of research, utilizing the FACS technology, which became a standard biology research tool that was part of numerous groundbreaking studies. Examples include Rolf Martin Zinkernagel (Nobel Prize in Physiology or Medicine in 1996, work circa 1985) [89] and Ralph M. Steinman (Nobel Prize in Physiology or Medicine in 2011, work circa 1989) [90] who relied on FACS for their Nobel Prize–winning research.

The time from the first proof of concept in 1970 to the Nobel Prize work enabled by the FACS technology was about 15 years. Prior to that work, though, the very first experiment with FACS was done by Herzenberg and Patricia Jones in 1974. They proved that individual B lymphocytes are committed to the synthesis of a single immunoglobulin heavy-chain isotype [91].

Before the emergence of the Coulter counter (1953) and FACS, cell biology relied on painstakingly slow staining of cells and counting under the microscope. Separation and isolation of cells by cell type was only done by dissection and—prior to the creation of laser capture microdissection—was inaccurate and time consuming. With laser capture microdissection, it is more accurate but still time consuming.

Coulter introduced the Coulter counter, in which a suspension of cells of interest—originally from circulating blood—is drawn through an electrically charged tube with a tiny hole at one end. The cell blocks the electrical field for a moment and is counted. The force and frequency of the distortion can be matched to specific types of cells based on their size.

Around 1970, the fluorescence-activated cell sorter was introduced and changed the way biologists and clinicians considered cellular research and clinical characterization. With this tool, they were able to achieve speeds of around 50,000 cells sorted per second based on molecular characteristics.

Zaffaroni, Read, and Pirrung—Microarrays

Breakthrough technology and invention rise from the creative efforts of engineers and scientists working together toward a goal. The creation of the microarray is a perfect example of this kind of process.

Affymax was founded by Alex Zaffaroni—also a founder of Syntex—and partner Avram Goldstein, M.D. and research pharmacologist, in order to develop automated drug discovery methods. Goldstein proposed peptide synthesis as the foundation of the effort, believing that short peptides could be ligands for cell-surface receptors. An important mission of Affymax at its founding was to build a large peptide library to be screened against various proteins in order to match peptides and their respective receptors.

At the time Affymax was formed, Mario Geysen had already developed a peptide synthesizer called Peptide on a Pin, but it had significant limitations. To overcome these limitations, Zaffaroni sponsored a series of brainstorming sessions to consider ways to generate the peptides in great numbers.

As the group discussed the goal, they had the notion that they would want to be able to synthesize not 600 peptides, but rather 6,000, or better yet 600,000 at a time-in parallel. This vast scale reminded one of the team members of the orthogonal technology of semiconductor fabrication.

J. Leighton Read—also an MD and a founder of Affymax—surfaced the idea of a semiconductor-like process to build the peptide arrays, and fortunately the group could draw upon Stanford faculty members (e.g., Fabian Pease, an electrical engineer specializing in photo lithography) who were familiar with the semiconductor fabrication process. Importantly, this familiarity with the fabrication process supported the feasibility of the idea, which then led to some offline work to conceptualize the process.

Later in an interview with the author, Leighton Read said that had they not had a group of diverse skills, "The idea may have fallen on fallow ground and never grown" [92]. Read and Michael C. Pirrung—a fourth Affymax founder—came up with a new approach using semiconductor fabrication methods and filed a patent in 1989. A Berkeley collaborator suggested that Stephen Fodor, a biochemist, join the company and lead the subsequent development effort [93].

After 18 months of Lubert Stryer and Stephen Fodor working on the concept, they published a 1991 *Science* article on a lithography-based peptide array. In their paper, they described how sequential deposition using semiconductor deposition process techniques could generate a precisely defined array of different peptides. The result of this development was that 1,024 peptides could be produced on a 1.6-cm^2 chip in a 10-step process. Although initially they had a 20-μm definition, higher densities could clearly be achieved, as the semiconductor industry was moving to submicrometer precision.

As the technology took shape, Fodor recognized that the process could be used to put down DNA fragments (oligonucleotides) in great numbers. The oligonucleotides would be used to capture target DNA from a solution. Then if a fluorescent marker specific to the targets is washed over the array, this can determine the number of target oligos present—a DNA diagnostic array. From the published work on the development of the microarray and based on an interview with Leighton Read, it becomes apparent that the development of the microarray was unlikely to have happened without the collaborative team of M.D./biologists (Read and Fodor), chemists (Pirrung and Stryer), and engineers (Fabian Pease, Peter Piekowsky, and James I. Winkler) [93]. Work began on the microarray, and in 1992 they formed a spinout called Affymetrix. Production of the gene chips started in 1994. The value of the microarray was enormous. With a chip and a reader, the user could detect specific sequences of DNA or RNA in a sample quickly, reliably, and with little effort. It revolutionized DNA testing and made it feasible for clinical applications and in forensics.

Since then, the microarray technology has been used in over 20,000 scientific papers [94] and no doubt will be the basis of important biology discoveries and recognition (possibly the Nobel Prize) in the future. Based on this book's historical perspective, one would expect that such work may have taken place around 2003 (9 years after proof of concept), and the Nobel Prize recognition would be around the time of the writing of this book (17.6 years after the work).

Common Features: Inventor, Multidisciplinary Team, Compressed Time Frame

It is important to conclude by noting that the technologies of this chapter share some common features:

- The technology was developed to address an unmet need in biology research.
- Engineering played a critical role in the success of these programs, though the engineers are generally not as publicly visible as the biologists on the team.
- For Sanger, Perutz, Moore and Stein, Oliver Smithies, and Andrew Fire, their time to a Nobel Prize was about 5 years from the first demonstration of the new technology they used. Sanger, Perutz, and Moore and Stein developed their own technologies concurrently with the research they conducted. Fire and Smithies used PCR in their work, which moved rapidly from Mullis's demonstration into the hands of numerous biologists. This is at least eight times faster than the average of 40 years for the other 31 Nobel Prizes this book described, where the researchers were obliged to wait until the technology was made available to the biology community.
- PCR, the DNA sequencer, FACS, and microarrays represent technologies born from unmet needs in biology research, and upon demonstration they were immediately put into service in biology laboratories around the world.

• Ten of the 11 research instruments discussed in this chapter needed to be commercialized before they were used in research; only PCR was used prior to commercialization.

There are alternative models for concurrent engineering. For example, it is possible to bring engineering technology into the research setting and develop the technology needed to do the research à la Leroy Hood or the LMB.

Another approach practiced successfully for decades in the physical sciences is for a community of researchers to identify a new enabling technology and to bring about the funding of a shared resource (such as telescopes, high-energy accelerators, etc.). The Consortium for Functional Glycomics is an example that is considered to be a success.

We should also like to emphasize the point that PCR, the DNA sequencer, FACS, and microarrays are examples of technologies developed expressly to meet a biological research need. The time from the initial demonstration of the key technology to its use in biological research was very short and in the case of PCR actually preceded the commercial introduction (e.g., Smithies). These technologies went from proof of concept to commercial introduction in an average of 5 years. This vastly outpaces the examples summarized in chapter 10 of genetic technologies between 1900 and 2000, which took an average of 40 years from concept to commercialization.

Another feature of these technologies is that while concepts were driven largely by biologists, engineers were critical to initial development. And in general, the work involved fairly big teams, well beyond a typical NIH grant, in order to integrate biologists, engineers, and scientists from other disciplines (chemistry, physics, etc.). For example, Affymetrix took advantage of the Small Business Innovation Research (SBIR) Awards at NIH and the Advanced Technology Programs of the National Institute of Standards and Technology (abolished by President Bush in 2007) and received nearly $38 million from just three contracts (two NIH and one ATP).

In general, the team size for these developments was larger than usual by biology research standards and thus required larger funding. The teams integrated biologist and engineers and were funded by at least one single large federal contract of over $20 million. While not at the scale of some of the high-energy physics projects, these projects are Big Biology by typical NIH standards.

Table 11.1 summarizes the characteristics of the projects described in this chapter.

Table 11.1
Summary of the projects described in this chapter

Technologies	Enabling Engineers	Initial Team Size/Funding	Proof of Concept	Commercial	Nobel Prize work	Nobel Prize
FACS	Mars project; Sandia		1970	1976 (BD)	Zinkernagel, 1985 Steinman, 1989	1996 2006
PCR	Atwood, Jones, Leath	20	1984	1987 (PE)	Smithies, 1988 Fire, 1991	2007 2006
Microarray	Pease (EE), Fiekowsky, Winkler	$38 million (NIH and ATP)	1991	1994 (Affy)	Forecast 1998	Forecast 2018
DNA sequencer		$33 million STC contract	1973 (Maxam and Gilbert) 1986 (Hood)	1987 (AB)		

AB, Applied Biosystems; Affy, Affymetrix; BD, Becton Dickinson; PE, Perkin-Elmer.

12 Institutions and Teams for Concurrent Biology and Engineering

The infrastructures and the teams that enabled the breakthroughs warrant more discussion than the mention in chapter 11. The individuals studied in the previous chapter are distinguished by their focus on technology (to varying degrees) and its application to their biology-centered research. Notably, they can also be linked by their common molecular biology research roots at one of three places: Caltech, the Rockefeller Institute for Medical Research, and the Laboratory of Molecular Biology (LMB) in Cambridge, England.

Therefore, it is relevant and important to take a moment to explore these institutions and their relative cultures that enabled such a valuable mix of technology and engineering. The author did not do an exhaustive search of the institutional origins of technology in Nobel Prize–winning biology. However, these three organizations seem to be at the heart of the molecular biology revolution and have enjoyed a rich history of technology development.

Caltech

For both Caltech and the LMB, their history of concurrent engineering appears to be tightly linked to two important and related factors. The first is each institution's efforts to increase exploration of molecular biology. The second is their ability to attract researchers with the skills and vision to become driving forces in the revolution of molecular biology.

As Phillip A. Sharp (MIT professor and Nobel laureate) notes, "Caltech said we want biology, we want the best biology, and we want genetic biology. They went and got Morgan in 1928 [Thomas Hunt Morgan, Nobel Prize in Physiology or Medicine in 1933]. So Morgan comes to Caltech, and Morgan sets genetics up at Caltech, and his department chair and dean protected genetics. So that's how they start."

> So [at Caltech] that was placed beside chemistry and physics, it was called biology at Caltech. There was some other biology, but that was biology.

And then Max Delbrück came in and Linus Pauling developed a new chemistry there. Linus created atomic chemistry; he took quantum mechanics and turned it into how you organize the teaching of chemical reactivity. Then he got interested in biological problems and went off and did the alpha helix and got his Nobel Prize.

As Sharp points out, this mixed-discipline culture of research created by Morgan at Caltech made it possible for highly valuable, synergistic research pairings to occur. For example, when Max Delbrück, a physicist, came to Caltech in 1937 with a Rockefeller fellowship, he was paired by Morgan with biologists such as Alfred Sturtevant to pursue genetics.

The culture of constantly pushing the boundaries of biological research—especially molecular biology—at Caltech also helped nurture the next generation of researchers. For example, William J. Dreyer, developer of the automated protein sequencer, mentored Leroy Hood, developer of the automated DNA sequencer.

An important aspect of this relationship, said Hood, was inculcating within him an ethos that has guided his career in biological research. "Indeed, Bill gave me the two dictums that have guided my subsequent scientific career," said Hood in a 2002 speech for the Kyoto Prize in Advanced Technologies [70]. "First, 'always practice biology at the leading edge.' It is more fun—always exciting and challenging. Second, 'If you really want to change biology, develop a new technology for pushing back the frontiers of biological knowledge.' A biologist should go wherever the biology takes them and always learn whatever new technologies are necessary for solving an unfolding biological problem."

It should also be pointed out that Hood was not the only researcher to benefit from the culture at Caltech. The vision to build a genetics—and thus a molecular biology—research program in the late 1930s, with the driving forces of Thomas Morgan, Max Delbrück, and Bill Dreyer, also led to the nurturing of the next generation of Phillip Sharp, Ron Davis, and later on Steve Quake. All of whom continue to use and develop technology to push biology discovery.

Laboratory of Molecular Biology

The Laboratory of Molecular Biology was started with Max Perutz as the lead and offers an instructive model of success.

LMB had its origins in the Cavendish Laboratory, part of the University of Cambridge. Cavendish scientists excelled in physics. J.J. Thomson discovered the electron there, and Ernest Rutherford smashed the atom. In 1915, Bragg, working with his father at the Cavendish, became the youngest person to win a Nobel Prize. Bragg (who wanted to see the melding of physics and biology) persuaded the Medical Research Council (MRC) to create the MRC Unit for Research on the Molecular Structure of Biological Systems in 1947. MRC's science budget grew about 17% annually. "Anything could be done. There were no limits," says Hugh Pelham, an LMB cell biologist. And to get those funds, all the researchers had to do was ask. After 30 years with MRC, Sir John Ernest Walker (Nobel Prize in Chemistry) says, "To this day, I have only ever written one grant." When LMB researchers needed a new instrument, Perutz made sure technicians and engineers were there to build it, a model he learned at the Cavendish. "We were interested in topics that stretched the techniques," says Walker, explaining how the laboratory developed technologies such as X-ray crystallography, DNA sequencing, and confocal microscopes.

Beyond the institution's many breakthroughs, which continue to this day, the LMB has produced a stunning number of Nobel laureates for the size of the faculty. In particular, the laboratory has won nine Nobel Prizes among 13 recipients, which have all been awarded for work conducted at the LMB on molecular biology. In addition, eight alumni have won Nobel Prizes, whose research, in some cases, was initiated and conducted at the LMB.

Of the nine Nobel Prizes won among the 13 researchers at the LMB, six of them were in chemistry and involved the development of technology to explore molecular biology. Of the remaining three, a brief read into the work reveals that novel technology development accompanied the scientific discovery.

Lastly, eight of the Nobel Prizes were awarded for work conducted at the LMB during the 30-year period between 1951 and 1980 (Sanger, Perutz, Watson and Crick, Klug, Milstein, Walker, and Sulston). The LMB was under the leadership of Max Perutz during this time.

It should also be noted that 1962 was a big year for the LMB. Watson and Crick as well as Max Perutz won Nobel Prizes, and the LMB was officially established as work began at its newly built Addenbrooke site.

Though the LMB moved to a new location from its early home in the Cavendish Laboratory, which itself was part of the University of Cambridge's Department of Physics, it remained true to its founding principles. These principles can be found in the constitution drafted by Perutz that established three divisions:

- Protein Crystallography: headed by John Kendrew
- Protein Chemistry: headed by Fred Sanger
- Molecular Genetics: headed by Francis Crick

The constitution was unique in that it created permanent expertise-centered divisions that were intended to live on past the retirement of their respective heads [40]. Another feature of the LMB was the cafeteria operated by Max's wife, Gisela. Coffee, lunch, and tea were served throughout the day, with the intent that all of the researchers would be pulled together in casual conversation. As Georgina Ferry writes in her biography of Perutz, "There was no rule that said you had to go to the canteen, but almost everybody did" [40].

The peer support among the great researchers at the LMB has been chronicled. Throughout the biography of Perutz, Ferry [40] describes the formal and informal interaction among the members. Critical thinking and questions among peers led to out-of-the-box thinking and discoveries.

Funding was also structured differently compared to a typical academic laboratory, which would have relied on a rather peripatetic livelihood of living from grant to grant [40]. Instead, money flowed from the Medical Research Council—an organization founded by the British government in 1920 to fund critical medical research—to the LMB through an annual budgeting process. This may have accorded the LMB greater flexibility to pursue long-term projects that involved concurrent engineering to support the molecular biology research.

A case in point, Perutz's initial characterization of hemoglobin provided the first picture of the structure. However, he wanted to continue his research in order to understand the functional changes to the structure when the heme group of hemoglobin bound and then released oxygen. To do this, he needed a very high resolution diffractometer.

A new instrument for Max's work was created by Uli Arndt. Arndt had been the chief instrument maker to Sir Lawrence Bragg and had worked directly with Bragg from 1939 to 1945—and throughout his career; Arndt continued to develop purpose-built technology to enable biological research.

Perutz was convinced that hemoglobin's ability to take up and give up oxygen was somehow governed by changes in the shape of critical areas of the huge molecule. The new diffractometer gave Max the first view into the two structural states of hemoglobin that he was looking for. He then was able to produce a series of papers that outlined the cooperative nature of some of hemoglobin's constituents in the presence of oxygen and their movement in a coordinated fashion upon binding to oxygen.

To take his research one step further and to demonstrate the intricacies of hemoglobin's mechanism, Perutz needed an instrument with an even higher resolution. This could only come from a much higher energy source of X-ray, which required moving the experiments to a synchrotron (see chapter 2) near Paris in order to get the images he needed.

In addition to the commitment to molecular biology (similar to Caltech), the LMB brought access to funding and had a culture of engineering and technology development. *One could probably see the machine shop just down the hall from the biology laboratory bench.* That was the nature of the organization.

The Rockefeller Institute for Medical Research (Now The Rockefeller University)

While the Rockefeller Institute has much notable and distinguishing scientific advancement to its credit, we have so far highlighted the work of William H. Stein and Stanford Moore. In particular, they demonstrated a propensity to develop enabling technology concurrently, which led to their incredible success exploring ribonuclease and deoxyribonuclease.

There is no doubt that the context and culture of the laboratory they worked in played an important role in their ultimate success. A driving force behind their work was Max Bergmann, who joined the institute only a few years before hiring Stein and Moore in 1937 and 1939, respectively.

Bergmann had been pursuing an understanding of the structure of proteins with Carl Niemann, and together they proposed a hypothesis of the repeating nature of proteins. Moore and Stein were consummate experimentalists and built the equipment and technology necessary to unravel the mystery of the structure of a protein.

Like the LMB, funding at the Rockefeller was often an internal matter. With a $690 million endowment (1950), the Rockefeller could afford to spend $30 million per year (5% of the endowment) on research in 1950, or the equivalent of $257 million per year in today's dollars.

Funding Constraints on Concurrent Science and Engineering

One wonders about the funding challenges that organizations face, as the scale of the efforts for concurrent science and engineering examples appears to require yearly funding of $15 million to $30 million per project, or nearly 10 times more than the typical National Institutes of Health (NIH) research grant. This is made all the more difficult as the Rockefeller Foundation is no longer funding basic biology research, leaving only the Howard Hughes Medical Institute (HHMI) ($16.1 billion endowment and $905 million/year for scientific research). However, the vast majority of HHMI funding goes to support individual investigators at approximately $1 million/year, thus it does not explicitly provide a source for funding concurrent biology and engineering projects. The historical precedence is captured in table 12.1.

Table 12.1
Estimated funding for selected projects

Technology	Project Funding	Source
Hemoglobin project (Max Perutz)	20 years, 20-person team	MRC and Rockefeller
DNA sequencer (Leroy Hood)	$15 million	National Science Foundation Science and Technology Centers (STC)
Electrophoresis (Arne Tiselius)		Rockefeller
Microarray	$30 million	National Institute of Standards and Technology (discontinued) and NIH Small Business Innovation Research
Genome sequencer	$1 billion	Largely DOE

Because these teams tended to be much larger compared to the current one or two principal investigator (PI) programs, the NIH historically has looked upon them as Big Biology. In particular, the NIH's principal (94% of yearly new grants) funding mechanism, the R01 grants, are typically only large enough ($436,000) [95] to support one or two principal investigators and a handful of graduate students as support staff.

The R01 program also predominantly responds to requests by researchers seeking funding for their specific areas of investigation rather than NIH articulating a need and receiving responses from researchers. This bottom-up approach has worked well in many areas. However, this focus misses opportunities where NIH could leverage its resources to identify promising avenues for research, articulate the need, and deliver funding in a directed manner.

There are other grant mechanisms (such as the so-called P30 Center Core Grants) that allow for larger teams and follow a more directed approach, but these are few and far between (new Center grants represent about 5% of all grants) [95]. Importantly, the R01 program is structured in such a way that it unintentionally favors projects that rely on off-the-shelf technology. Researchers submitting a grant proposal for R01 funding are required to have preliminary data to demonstrate the viability of their planned research. This creates something of the chicken-or-the-egg scenario.

Imagine that you are thinking of going down a new research pathway that requires the use of a novel technology, which you wish to develop. But you are required to submit preliminary data. How can you create the preliminary data without the novel technology? To submit an R01 and meet the preliminary data requirement, you already have to have developed the technology and proved its efficacy before you receive funding. Unfortunately, creating a new technology can be the most expensive piece of the puzzle you are trying to solve. Who is going to pay for it?

The NIH does have an R21 program, which does not require preliminary data. However, the award size ($217,000) [95] is a fraction of that of the R01 grants, which means you simply cannot develop technology with an R21.

As Eric Lander, an MIT professor of biology and a leader in the Human Genome Project, noted in an interview for this book, "It was lucky that the Genome Project had a funding path and the Broad Institute was largely built out of that. It was built out of one project, where one could bring together these multidisciplinary teams. Some funding has continued for that, but biology needs a lot more of it because as we move to things that are more practical and more useful, there are still academic projects. Most of what we have to do are not commercial projects, such as find all the mutations that drive a cancer."

What this all points to is the tension between NIH's funding process, the politics and concerns over Big Science, and the vast potential for directed programs using concurrent engineering to support research. The predominant funding model used by NIH, while good at many things, is constructed in a manner where concurrent technology development is precluded, and perhaps significant opportunities are being missed. The dilemma of incorporating new technology created by NIH's process should be addressed.

Big Teams and Big Science

"This book is about the value of parallel biology/engineering efforts," observed Dr. Jeffrey Drazen, editor in chief of the *New England Journal of Medicine*, and a contributor to this book. "This can be small—a one- or two-person shop with the needed bio and physics background *or big* with many people of varying backgrounds."

Big can be quite big. As discussed in chapter 2, science is enabled by "big" technology like cyclotrons or the Hubble Space Telescope, which opens the discussion of "Big Science."

In 1961, Alvin M. Weinberg, director of Oak Ridge National Laboratory, set the stage for much of the debates to follow when he published his rather prescient paper on Big Science [96]. As one of the first to coin the term, Weinberg referred to the development of high-energy accelerators as Big Science and compared the twentieth-century development of similar technologies to the pyramids, the Sphinx, Notre Dame of Paris, and Versailles.

In one instance of rhetorical flourish he writes, "When history looks at the 20th Century, she will see science and technology as its theme; she will find in the monuments of Big Science—the huge rockets, the high-energy accelerators, the high-flux research reactors—symbols of our time just as surely as she finds Notre Dame a symbol of the Middle Ages" [96].

Notably, although Weinberg's paper may have launched the continuing debate about Big Science, he discusses the importance and, to some degree, the inevitability of Big Science. In a completely overlooked point of his paper, he lobbies for the application of Big Science to biology: "I suspect that most Americans would prefer to belong to the society which first gave the world a cure for cancer than to the society which put the first astronaut on Mars" [96]. Albert Sabin [97] agreed with Weinberg calling for a "concentrated attack on the more complex disease problems and the need to develop the necessary mechanism to facilitate and coordinate collaborative research." The author believes that for some complex biology problems, which would benefit from a large collaborative and concurrent engineering biology project, Big Biology is called for.

Big Biology

The author suggests that the prototype of Big Biology was the project to discover the structure of hemoglobin by Max Perutz or the first sequencing by Sanger. Going forward in the future, this book will identify in the next chapters several technologies that would represent Big Biology:

a. reveal real-time in vivo cell signaling (both intracellular and intercellular) quantitatively for numerous pathways;

b. detection and localization of tumors prior to vascularization (less than 1 mm);

c. ex vivo organ models that recapitulate human systems (including the immune response).

The Human Genome Project (HGP) is often referred to as the "prototype of Big Biology" [98], but it may not be the best example. Although it involved concurrent engineering (see the discussion with Eric Lander in chapter 11), around 1985, David Baltimore (Nobel laureate), then of the Whitehead Institute, said, " I don't see the lack of the sequence of the human genome as limiting factor in anyone's research. . . . you can easily get the sequence once you've identified the piece of DNA you're interested in" [99]. Baltimore is pointing out that researchers could sequence DNA in their laboratories prior to the development of the technology that was part of the HGP. The HGP did not provide unique data or information that could not otherwise have been obtained (though at huge cost in time and money).

While the HGP may not be the best prototype for Big Biology because sequences could be done in many laboratories, still the story of the HGP is important to tell as it reveals some of the difficulties that arise when attempting to undertake a large directed project in biology, which involves significant engineering of technology.

The story surprisingly starts with the Atomic Energy Commission (AEC) because its predecessor, the Department of Energy, was the sponsor of the HGP. Started in 1947, the AEC fostered much of the development of the Big Science of high-energy physics and helped create the Lawrence Radiation Laboratory (see chapter 2) as a national laboratory. Beginning in the mid-1970s, the AEC went through organizational changes that culminated in its duties being incorporated within the Department of Energy.

Since its inception in 1947, the AEC has sponsored the apparatus development and subsequent basic research of the high-energy physics community. Throughout, the AEC developed the infrastructure and culture to support programs where significant engineering was required to build leading-edge apparatuses in order to support scientific research.

Thus in 1987, it was of little surprise that biologists, who advocated sequencing of the human genome, turned to the Department of Energy (DOE) for financial support. So far in their journey, they had been unsuccessful in getting support from the NIH for this rather large effort, which represented a mix of biology research and engineering.

In April 1987, the "Report on the Human Genome Initiative" was created at the Lawrence Berkeley Laboratories for the DOE's Office of Health and Environmental Research. The report laid out the objectives and funding—$200 million per year—for what became the Human Genome Project. In the cover letter the authors say:

> It may seem audacious to ask the DOE to spearhead such a biological revolution, but scientists of many persuasions on the subcommittee and on HERAC agree When done properly, the effort will be interagency and international in scope; but it must have strong central control, a base akin to the National Laboratories, and flexible ways to access a huge array of university and industrial partners. We believe this can and should be done, and that DOE is the one to do it. [100]

Ultimately, the genome project was a joint program with a wide range of participants, and the NIH funded the lion's share of the effort: $2 billion of the total $2.7 billion through 2000 [101].

Still, the initiative came from the DOE, and the statement above is profoundly important for understanding the convergence at the heart of this book. On one hand, scientists (either physicists or biologists) must articulate the science goals in terms that the engineering and science community can understand, and on the other hand, the execution is an engineering problem and should be funded and managed by people that are experienced in successfully managing large-scale projects.

As we look at some of the big challenges in biology and life sciences in the post-genomic period (chapters 15 and 16), we need to ask ourselves what type of funding and co-sponsorship should these projects have in order to succeed? This question is all the more critical when there is so much dependent on the successful execution of the development of the apparatus to support the basic research.

Does NIH Need a DARPA?

Professor Steve Quake (see chapter 11) mentions the Defense Advanced Research Projects Agency (DARPA) model for funding advanced technology development. Given the experience of the HGP and the need for an organization comfortable with Big Science to pitch in, it is well worth asking if NIH needs its own DARPA.

In a provocative article entitled "Does NIH Need a DARPA?" [102], Robert Mullen Cook-Deegan, a physician and former government technology assessment expert, offers a starting point. In his article, Cook-Deegan spotlights the inherent limitations of NIH's peer review process. He argues it is overly conservative and prone toward, "The safe squeezing out [the] novel."

By contrast, he writes, "In DARPA culture, managers are self-avowed scientific and technological fanatics. Their base skill is recognizing talent that is relevant to defense needs and providing funds for its expression. The DARPA institutional ethos is described "as 80 decision makers linked by a travel office," which emphasizes its highly interactive (at times intrusive) style. It is ironic that within one of the world's most notorious bureaucracies, the Department of Defense, resides a tribe of rambunctious technological entrepreneurs."

DARPA recognizes that investments in projects, where new technology must be developed or existing technology put to a novel use, can lead to incredibly valuable results. However, rather than wait for such proposals to percolate up to them, DARPA seeks out critical unmet needs where this type of funding and research model could be put to best use.

Cook-Deegan advocates for an incremental approach led by a pilot test followed by limiting this approach to only a few select projects. In fact, he sees considerable risk in turning NIH entirely over to a DARPA-like model.

Cook-Deegan writes, "Much of the most important work supported by NIH and NSF is conducted through tens of thousands of relatively small grants. Innovation bubbles up in unexpected places thanks to the flexibility of the grant mechanism, which leaves funds largely under the control of investigators. NIH handles 45,000 grant applications per year. It would be folly to adopt DARPA's methods for so many small projects covering enormous areas of science."

While this approach makes sense and should be tried, it is also important to recognize that to date, NIH has a very different culture from DARPA. This cultural difference between the two organizations, in and of itself, could hamper execution of a proper pilot program as well as adoption of an effective system of selectively seeking out promising opportunities.

Instead, NIH could commit a block of funding, placed within a temporary limited-life interagency program, where DARPA professionals could help place that funding. Rather than attempting to create a new institution from scratch and working through the learning curve of developing directed research, the culture of DARPA could be grafted into that of NIH and then ungrafted once the program is running well.

This does not have to be a permanent situation either. In order for the graft to take root, it could likely be done for two funding cycles, say 6 to 8 years, whereby the program would take root and create its home within the firmament of the NIH.

Given the incredible potential of concurrent engineering, such a directed funding mechanism could lead to timelier and greater breakthroughs.

Cultural and Operational Challenges of Concurrent Engineering

Concurrent engineering has an important role to play within the world of academic research institutions. However, there are challenges to be overcome.

To gain some insights into the challenges of concurrent engineering and biology science within academia, the author interviewed Linda Griffith, a professor within the MIT Biological Engineering Department. Linda discussed the environment for collaboration on the large interdisciplinary projects inherent in concurrent engineering.

> In engineering there has been more acceptance that people collaborate and when you go up for tenure there's other ways that you can parse who did what.
>
> So I think culturally it seems in biology that the promotion review process—and again this is my impression from the outside—everything is supposed to be your own. That includes you get an R01 and you're the PI. If you don't have an R01 or multiple R01's by the time you go for tenure, you're viewed as not successful.

Whereas in engineering, it's completely possible to get tenure being a co-investigator on multi-investigator grants in which other senior people are PI's. That's done all the time at the school of engineering.

The author asked Matt Mountain, director of the Space Telescope Science Institute, about the institutional challenges integrating astrophysics research with the engineering development of the Hubble Space Telescope.

And when we have complex systems, the only thing that really makes a difference is somebody must have the full scientific knowledge of what's required that is the science champion. You have to have, on any of these successful projects, a really strong science champion or champion team, depending on the extent of the program, who really always goes back to the science and let's work this back through and let's work at how to trade these things off in a sensible space.

I think the important thing is that Rodger [Dr. Rodger Doxsey; see chapter 2] had the credibility on the engineering and the science side. You only get these complicated projects going forward when you get consensus, which is distinctly different from compromise, by the way, consensus. Rodger was able to generate consensus across the science and engineering divide. It's not the same as compromise. But Rodger was the key to getting both sides to move forward, which you have to do in these complex teams. When anybody ever says to you the science leads or the engineering leads, neither of that is true.

Jeffery Drazen

Dr. Jeffrey Drazen is editor in chief of the *New England Journal of Medicine*, a physician, and a researcher in pulmonary medicine. For full disclosure, Drazen served as a primary advisor for this book and was instrumental in suggesting the topic and overall direction. During summary discussion with Drazen, he provided more clinical perspective on the integration of engineering and biology research that the other interviews did not quite cover, and the author decided to include it in the book.

"I have always considered myself an engineer really," said Drazen, "because when I went into this business one of my advisors said how could you go into a field that didn't have its own set of Maxwell's equations. [Biology] is very soft and a lot of biology is done on syllogistic reasoning where if this happens that happens and therefore that must happen. It is not quantitative and I like things that are generally quantitative as opposed to qualitative."

Technology in the Biology Laboratory

Early in our interview, Drazen offered a story of how his own engineering acumen helped speed and make more efficient the research of one of the first laboratories he worked in.

I went to respiration. We were interested in asthma and what could cause airway constriction. A substance had been discovered, this is the biology side of it, in the late 1930s that caused constriction of a guinea pig GI smooth muscle. So I show up after having decided to pursue breathing because I took a course in my second year that was taught by a guy named Jere Mead, who was a respiration physiologist, and John Pappenheimer, who was also a respiration physiologist, but it was very quantitatively oriented and I liked it.

I thought I wanted to be a respiration physiologist and Jere Mead told me I didn't want to be one. He said I wanted to learn respiration physiology and use it to understand the biology of disease. He said that because we did not understand many clinical problems that are manifest as physiological abnormalities.

So I went to work and he told me, this is in 1969, that he thought the two big problems were the control of breathing during sleep and asthma. As I had had a patient with asthma assigned to me when I was a first-year medical student, it was clear that we knew pretty much nothing about asthma and asthma treatment.

So I went to work with Frank Austen. I showed up in his lab in 1970. To record the constriction of a guinea pig ileum they would kill a guinea pig and then take out this little piece of smooth muscle about an inch long and they would hook it up to a lever system. One end of the lever was connected to a stylus that was writing on a piece of paper that had been smoked.

I said to Frank Austen, you guys are back in the dark ages. We have ways of recording this electronically. I bought pressure transducers and electronic recorders and put [Austen's laboratory] into the electronic recording business. It was interesting that I was the only guy in the lab that understood it. You had both the gain and the span and you had to balance these transducers so their working range was within the page. It was basically a Wheatstone bridge but people were totally dumbfounded by it.

But as far as people were concerned it didn't advance the biology. All we were doing was replacing a piece of old-fashioned technology that was way behind the times.

That taught me that the people who were interested in biology didn't care about instruments. They had all of these instruments in the lab but they didn't understand how they worked. They just saw them as measuring devices.

For many people that was all they cared about. They did some biology and they wanted to make a measurement of it. They needed a way to quantitate something they were doing, whatever was commercially available that they could use.

Asked about how to push the research frontiers, Drazen answered, "You need to have technologists work with you to push it."

The other part of my time I was working in Jere Mead's lab. He was a physiologist so he was measuring pressures and flows; those were the things that he cared about. The people in Frank Austen's lab were interested in the rates of enzymatic reactions or the fact that cells would produce A, B, and C, but in Jere Mead's lab the primary variables were pressure, flow, and volume.

So he had engineers working there who kept us up with the latest technology to measure pressure, flow, and volume so we were able to do experiments that nobody else in the world could do.

In Eric Lander's shop, they have some guys who are sequencing bases and they are pushing new technology.[1] With this group of people they were able to do a base pair for a fraction of a penny as opposed to a buck and a half as it was ten years ago. Other people can focus entirely on the biology end of it.

So you need to have somebody, such as with Perutz, who is running the lab and is smart enough to figure out where the problems are. At the Laboratory for Molecular Biology it was X-ray crystallography where they had all those guys making X-ray tubes and reading the diffraction and so on.

Funding

The discussion then turned to one of the big challenges described in this book: cancer pathway discovery and drug development. Drazen picks up the conversation with the nature of the research and challenges:

So clinical trials are all big experiments that we are learning from. For example, you can test a pathway inhibitor in mice, but in humans things often don't work the way you expect them to because the system is much more complex in man than in mice. There is always another path and [to succeed] you have to know what the other path is.

The problem with our current approach is that it is siloed. Either you are a path A guy or a path B person. If you think too big, with too many pathways and your grant is reviewed by your peers who are all one-pathway people, your proposal is viewed as diffuse and you don't get funded; you are out of business.

There is also the issue that combining research for multiple paths entails much larger projects and much larger funding. As Drazen noted, R01 grants, the primary NIH funding mechanism, often are not enough to bring these larger projects to fruition.

To fund bigger and riskier projects, the guy running it [funding agency], he or she has to be willing to take a number of hits on things that don't work. This is what venture capitalists face all the time. Venture capitalists don't get into things until there is a commercial sign. So this is like venture capitalism except you don't make money, you make ideas and you have to be willing to take a chance on something that seems to be useful and the profit isn't going to be $100 million, it is going to be ideas.

This could be structured like a Longitude Prize [a prize established by the British Board of Longitude for the person who could invent a means of determining longitude at sea], but then somebody has to define the problem.

For example, think about the problem that has come out recently. Up until I would say a year ago people considered the genome to be a lot of junk DNA, but then more than four or five years ago they said well it can't be junk. It has to have some function. And we are beginning to realize that it is probably regulatory rather than junk.

And then suddenly there are all these grants and all this work on that project so it is a very biological thing. I don't know if it took new physical stuff. There was sort of big biology because here was the conundrum of what's all that doing out there and someone said it must be useful. And then enough experiments were done by enough different people to show that this stuff is useful.

The New Technology, the New Fact, and the Paradigm Change

My interview with Drazen came after he had the opportunity to read a draft manuscript of this book. As such, his comments were informed by the material presented in the book. This allowed Drazen to offer some compelling commentary on the process and method by which concurrent engineering could be most effectively executed.

I looked at this book as saying the really big progress in biology comes when there is somebody who is visionary enough to say, "If I can measure something," and then he is able to work with somebody who [provides the expertise] to help measure it. Then they are able to get into an area where no one has ever been before and as soon as people realize how important this is, the instrument companies get into the business of making it.

It's the challenge for the person who doesn't even know how to ask the question. I don't think we have thought enough about how to even ask the question. All of our thinking is you want to be an immunologist so as a graduate student you spend four years learning all of the facts and the kinds of experiments from which the facts were derived. Now it is your turn to do experiments and you do experiments just like the ones from which the current facts arrived in order to derive a new fact.

Reflecting on the challenge, Drazen said, "There needs to be a way for people to rub together. You know this Bell Labs idea. One of the things that struck me, from the mid-1980s to the mid-1990s, if you went to a lab that wasn't crowded and each person was working in their own space with their own stuff; you knew this was a place you never want to be. It would be really boring because you want to be in a place that is crowded enough so that people who wouldn't ordinarily interact with each other interact with each other."

For example, at the Brigham we were short on space when I moved over there in the early nineties. By luck, I ended up next to a mouse geneticist named David Bier. Suddenly he had problems that he never thought were able to be studied and suddenly I had ways to study problems I never thought existed. We did a bunch of mouse crosses and we ended up working with Eric Lander to do the genotyping, we made terrific progress. This only happened because an administrator parked our labs next to each other and we happened to talk about science at the lunch table.

Finally in thinking about what made the Laboratory of Molecular Biology so successful in the 1960s and 1970s, Drazen asked: "Who is it that made Max Perutz successful? It had to be somebody that was willing to pay the bills. There had to be long periods where things didn't work. He had to have some friend somewhere who was writing the checks. He needed a John Beresford Tipton, Jr. [the unseen philanthropist of the "Millionaire" television show].

Drazen's question can be answered by Perutz's own words [81], "Sir Harold Himsworth, as secretary of the Medical Research Council, made sure that we never lacked any of the funds or equipment needed for our work. Without his consistent support, research on the required scale could not have been done. Dr. G. R. Pomerat, as one of the directors of the Rockefeller Foundation, helped us to cover any additional expenditure, which the Medical Research Council had difficulty in meeting. Moreover, the Rockefeller Foundation supported me in the early years; without this support, the work would probably not have got beyond its initial stages."

Drazen added, "That's what needs to happen."

13 Concurrent Engineering in the Clinic

Thus far, we have examined the effects, or in some cases lack of effects, of concurrent engineering in the biologist's lab. However, this is not the only medical area within which concurrent engineering has played a significant enabling role.

We would be remiss if we did not take a look, albeit relatively brief, at the impact of concurrent engineering in the clinic. In particular, this chapter examines the role concurrent engineering played in developing a handful of the imaging, diagnostic, and therapeutic technologies in use in medicine today.

This chapter is distinct from the previous chapters about the technologies that enabled biological discovery because the end point is significantly different. The end point of these stories is the release of the technology to the commercial world and entry into the clinic while the end point of the prior case histories is biological discovery. Clinical devices nearly always involve the integration of engineering and medical practice or biology. We did not include these studies in the timeline analysis from proof of concept to commercial clinical product (see chapter 10). The purpose of these case studies is to provide a contrast to the development stories that led to biological discoveries.

That said, Prof. Annetine Gelijns of the Icahn School of Medicine at Mount Sinai observed, "Although these clinical technologies may not exhibit the same time profile as the basic biological discoveries [the] overall hypothesis still holds and concurrent engineering, that is, the effective coupling of engineering to clinical medicine, speeds the introduction of new diagnostic and therapeutic interventions" [103].

This will not be the first tour through the interface of engineering and its use to develop tools for physicians to use. There have been a number of studies and books on the topic of engineering in the support of life sciences and/or medicine.

A recent book by Andras Gedeon [104] highlights 99 major contributions of science and technology to the advancement of medicine over five centuries.

Another treatise by Dr. Linda Griffith and Dr. Alan Grodzinsky [105] on bioengineering lists an additional 20 contributions. These works, particularly the Gedeon book, which contains over 1,100 illustrations, provide a wonderful sweep of the history of medical advancement via the contributions of science and technology.

As such, these works provide a backdrop to the question: How did the significant advancements come about? More specifically, was the chasm between biology and engineering bridged by a single man or by a team of collaborators with different skill sets? If both situations existed, which achieved the necessary breakthrough in a timely fashion?

Parenthetically, from a brief look at each case, this book also attempts to draw out key insights as to the critical elements of its success.

MRI and Its Beginnings in NMR

The origins of magnetic resonance imaging (MRI) are not without controversy. In 2003, the Nobel Prize in Physiology or Medicine was awarded to Dr. Paul Lauterbur and Sir Peter Mansfield "for their discoveries concerning magnetic resonance imaging." However, awarding of the prize to these two men omitted the critical contribution in the development of MRI by Dr. Raymond Vahan Damadian.

Therefore, many believe the important contributions of Damadian require that he also be recognized by the Nobel committee.

While this controversy dominates much of the history of MRI, there is another dimension of the Damadian and Lauterbur dichotomy that is of interest to this book. An important clinical technology, MRI grew out of nuclear magnetic resonance (NMR) technology, which has no clinical presence and was used only for research. However, NMR technology and experiments from 1920 and into the 1950s were the forerunner to MRI and paced the progress toward medical applications.

The "M" in NMR and MRI stands for *magnetic*, and both technologies require intense magnetic fields in order to extract the weak signals from the samples they are observing. The fundamentals of NMR date back to the beginning of the twentieth century with Joseph Larmor's findings that describe the frequency of precession of a nucleus. Nuclei of atoms naturally rotate, and their inherent electric charge causes a small magnetic field of their own. When an outside intense magnetic field is applied to a material, the field will not align perfectly with all of the spins of the material's nuclei and will cause the naturally rotating nuclei to wobble. This has since come to be known as the Larmor precession.

Then in 1938, Isidor Rabi demonstrated resonance peak of lithium and chlorine with a small, single-purpose NMR unit.[1] The Rabi experiment showed the resonance of a gaseous molecule, but it was not designed to test other molecules.

Fortunately, Rabi's work tipped off several physics investigators at Stanford, namely Dr. Felix Bloch and William Webster Hansen, to explore the practical use of nuclear resonance for chemical analysis.

Through the 1940s, the Bloch and Hansen laboratories worked on characterizing numerous molecules with NMR. Similar work was taking place at Harvard by Edward Purcell using a single radio-frequency coil while Bloch at Stanford used a dual-coil approach. Bloch developed the macroscopic relations governing nuclear magnetization in an external magnetic field. This carried the work of Rabi into a more complete understanding of the governing relations, which allowed NMR to be a practical tool.

Bloch and Hansen drew a number of young Stanford graduates into their work, including Russell Varian, Martin E Packard, and W.G. Proctor. In particular, Bloch was supported by Varian, who acted as a laboratory assistant in the physics department at Stanford. There is little doubt that Varian was a very adept physicist in his own right as he is credited with designing the magnet used in the Stanford labs to study NMR.

In total, the Bloch team managed to transition the concept of NMR into a working unit with analytical capabilities.

Like numerous other Stanford graduates to follow, Varian, Packard, and Proctor were entrepreneurial in spirit and formed a company called Varian Associates in 1948 (figure 13.1). Among its many successes, the company

Figure 13.1
(Left) Varian "headquarters" in 1949 in San Carlos, California. (Right) Russell Varian and Martin Packard with an early NMR instrument in the first applications laboratory in Building 1.
Source: Communications and Power Industries (CPI) documents of Varian Associates. Reprinted with the permission of CPI.

developed the first NMR unit in 1952, the Varian HR-30—a room-sized instrument—which was installed at the Humble Oil Company. As with thousands of other laboratories, the oil company needed to understand the composition of the crude and refined products, and NMR became the analytic tool of choice for many of the organic chemists.

Through the 1950s and 1960s, continuous improvements in the NMR instruments were made, which significantly reduced the size and the cost. Additionally, numerous units were placed in research units throughout the country.

One such unit was at the State University of New York (SUNY) Downstate Medical Center in Brooklyn, New York, where Dr. Raymond Damadian arrived in 1967 to study sodium and potassium characteristics of cells. It was his hope to explore whether NMR could distinguish one cell type from another.

Observing that cells exhibited different NMR signatures, Damadian decided to test normal tissue versus cancerous tissue, which led him to observe a difference in NMR signature. He published the results in his now famous *Science* article in 1971 and filed a patent on the use of NMR to find cancer in humans.

As such, Damadian transitioned the NMR technology to its use in biology and medicine, an important step toward MRI.

Shortly after the 1971 publication of the detection of resected cancer tissue via NMR, others in the field began testing the bounds of NMR. In particular, I.D. Weisman, L.H. Bennett, L.R. Maxwell, M.W. Woods, and D. Burk demonstrated detection of cancerous tumor growth in live mice using NMR. They further speculated in their 1972 *Science* article that NMR could be used as a nonsurgical technique for cancer detection.

Attention now turns to Paul Lauterbur, also working within the SUNY system, though at Stony Brook. Lauterbur, a chemist, developed an interest in material characterization while in the U.S. Army, where he experimented with NMR to detect trace amounts of specific chemicals in a sample.

After leaving the military, he returned to the Mellon Institute of Industrial Research in Pittsburgh where he worked on his Ph.D. in silicon and carbon NMR spectroscopy. Before leaving Mellon, he helped set up NMR Specialties in western Pennsylvania. Then in 1963, he moved to SUNY Stony Brook and continued his NMR research.

In 1971, he had the idea that magnetic gradients could be used to create a spatially defined image and began experiments with pairs of capillary tubes filled with water. This was a truly brilliant insight and deserving of the recognition he received. The basic relation governing the NMR state is that the frequency of the NMR signal heard by the instrument changes with the intensity of the magnetic field applied to it. The magnetic field is not perfectly uniform spatially, and much work went into "shimming" the magnetics to make the field uniform. Lauterbur turned this weakness into strength. He realized that by intentionally varying the magnetic field across the specimen in a controlled way, the radio-signal frequency from the NMR signal would vary in a precise way by specific location, and therefore, the emitted radio signal from the sample could be associated with its location limited only by the precision of the magnetic gradient and by the sensitivity of the receiving device. By linking the NMR signal to its three-dimensional location, one could reconstruct an "image" of the specimen, and hence NMR (which does not slice the specimen signal into discrete locations) became MRI. Today, the resolution of large clinical MRI instruments is about 2 mm, whereas research MRI tools can even focus on a single cell.

In 1973, Lauterbur published his results and thus began the development of MRI technology.

The major NMR and MRI events are summarized graphically in figure 13.2.

The guiding principles of magnetic resonance were developed abstractly (e.g., theorem) by Larmor around the beginning of the twentieth century. Then Isidor Rabi set out to demonstrate the validity of the Larmor theorem through an exquisitely designed experiment that showed a precise magnetic-created resonance of the nucleus could be measured.

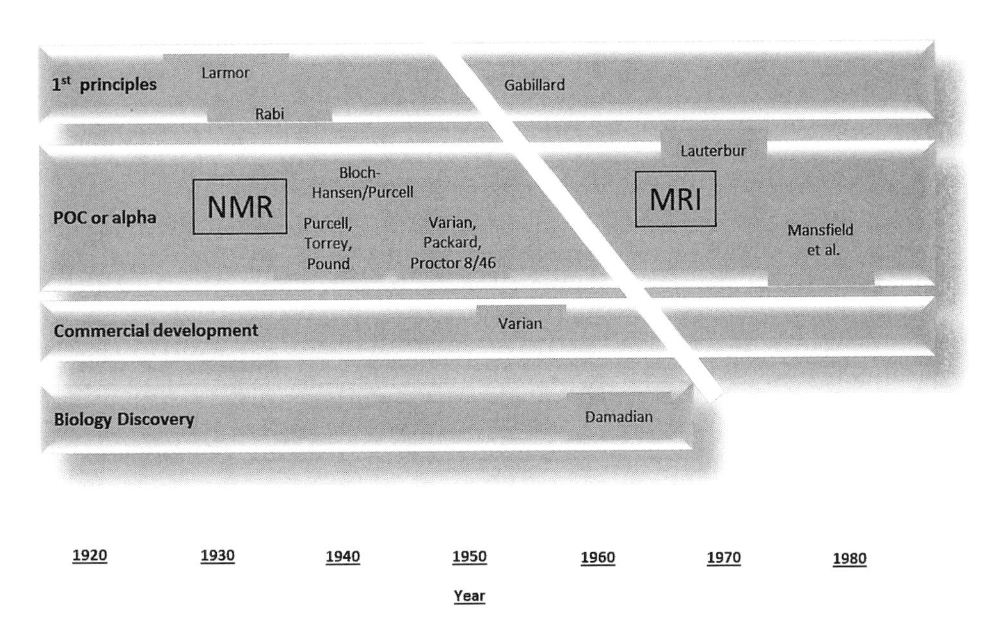

Figure 13.2
The NMR and MRI development timelines and the people that made it happen.

Rabi's 1938 experiment represents the demonstration of the first principles of NMR. However, subsequent proof of concept (POC)—the application of scientific and mathematical principles to practical ends—lay within the realm of engineering. As such, Rabi's device was designed specifically to confirm the NMR of a specific gas and could not be used to characterize materials and certainly not solids. No attempt was made to devise an instrument capable of measuring the parameters of interest with any other molecule or phase of material besides gas. That POC had to wait another 20 years, until 1946 when Bloch and Varian's teams at Stanford engineered the concept into operational instruments.

Unfortunately for the practice of medicine, the NMR instrument was not made readily available to the scientific community until after 1953 and only then for its use in "applications to chemistry" [106].

Bloch, Varian, Proctor, and Packard engineered the first NMR instrument (a room-sized NMR unit for petrochemical analysis) in 1946, but it was not used in any other laboratory until 1955 and not until 1971 in a biology laboratory (Damadian). Why did it take so long?

The simple answer is that others could not reproduce the technology easily from conventional laboratory equipment. This meant that distribution to the biology community had to wait for engineering (size and cost reduction) to occur through the commercialization process, which added another 16 years onto the 10 years from the POC. Therefore, the total timeline for this important technology was 26 years.

An NMR Instrument, 1950

Felix Bloch, the Stanford physicist working with William Webster Hansen, set out to study NMR and its use on a variety of materials in order to develop an understanding of the governing relations for NMR spectrographs of materials. Ultimately, the two men developed what has become known as the Bloch equations and related materials characterization techniques.

Meanwhile, Varian, Packard, and Proctor, while at the Department of Physics at Stanford, were tasked with developing a more reliable instrument. Their papers and Packard's Ph.D. thesis manifest the electrical and mechanical engineering that was involved in developing the first operational NMR unit in the laboratory to support the research of Bloch and Hansen.

Engineering the technology to support physics research is not an uncommon thing in physics. The two disciplines speak the same language, and it is commonly recognized in physics that to explore new physical phenomena generally requires new instruments. The tight integration of science and the enabling engineering is a natural process in the physical sciences, as has been pointed out previously in this book.

Commercial NMR Instruments, 1950–1970

Through the 1950s, commercial instruments were developed, most notably by Varian, but NMR Specialties struggled for years as demand for NMR instruments was uneven, and the company was often near bankruptcy. It wasn't until 1961 that a "bench-top" spectrometer designed for the laboratory chemist was offered by Varian. The spectrometers slowly spread and became available to researchers in academia.

Medical Breakthroughs, 1970s

While at SUNY Downstate Medical Center in Brooklyn, New York, Raymond Damadian developed the concept whereby he believed cancerous tissue could be distinguished in vivo via NMR and set out to prove his hypothesis. A commercial unit was available on campus, and he arranged time on the instrument. By 1971, he had found a measurable NMR signal difference between cancer tissue and normal tissue. In the long run, however, it was shown that these differences were not unique to cancer and that although NMR could distinguish various tissues, it was not a sensitive and specific method to distinguish normal versus malignant tissues.

As a result, his work tripped off a number of investigations seeking to reproduce his results. This, in turn, sparked a larger drive to perform a wide range of in vivo imaging with NMR.

An Alternative History: Accelerated Development

The 1940 Stanford team—Bloch, Hansen, Varian, Proctor, and Packard—was a seamless integration of scientists and engineers who embarked on an exploration of the capacity of NMR to characterize molecules. The engineers—Varian, Proctor, and Packard—are known to us because they went on to create Varian Associates (now Varian), a large, successful company that maintained the history and the engineering contributions of the founding employees.

While probably not recognized at the time, these three researchers represent a multidisciplinary team. Bloch, the theoretical physicist, engaged the talents of applied physicists (engineers) to create the instrument he needed to explore the science.

However, it is possible that the Bloch, Hansen, Varian, Packard, and Proctor team could have accommodated a collaborator from the Stanford University School of Medicine. This addition might have performed the ex

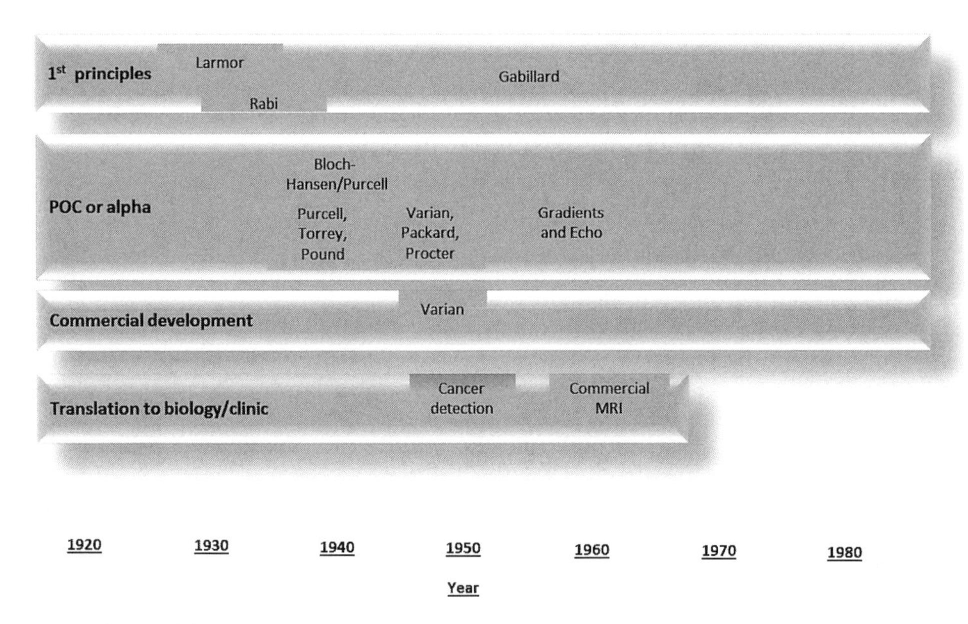

Figure 13.3
An alternative history with accelerated MRI commercialization assuming Varian had experimented with NMR in human tissue in 1950 and created the motive to apply NMR to medical applications. This might have propelled the translation of Gabillard's one-dimensional spatial resolution work into three dimensions creating the first MRI in 1960 rather than 1970.

vivo experiments (completed by Damadian in 1971) in 1948, thereby accelerating the biomedical application of NMR by 23 years.

Instead, Damadian's crucial experiment had to wait 23 years for the NMR technology to move into commercial production and become readily available. Achieving the translation of NMR to biology in 1950 seems eminently possible as shown in the accelerated view of the NMR and MRI development in figure 13.3.

The question remains whether the advancement of MRI and the fulfillment of the medical imaging breakthrough could have been accelerated as well. The physical principles of magnetic gradient and shimming were published by Roger Gabillard [107] in 1951. It is also possible that the concept of developing spatial resolution with magnetic gradients, first performed by Lauterbur in 1971, could have been proposed in 1950.

However, could the engineering of the MRI, as we know it, have taken place in the 1950s? It would appear that with adequate focus and funding, a commercial MRI could have been accelerated as the guiding principles had already been published. The electronics (albeit nonintegrated circuit) for low-resolution imaging could have been done in the 1960s. Additionally, superconducting magnets were already being used in NMR prior to 1960. Moreover, Johann Rodan published the mathematical framework for image reconstruction in 1917. Allan Cormack then published the mathematical analysis to tomographic image reconstruction in 1963. The computational requirements of the modern MRI, however, were not widely available in 1960. Nonetheless, that would not have stopped someone like Perutz, who, as previously discussed, built the EDSAC II computer in 1958 to do the complex analysis of the structure of hemoglobin from X-ray diffraction patterns. Certainly, one could find that kind of computational talent at Stanford in a nearby building.

Ultrasound

As with MRI, ultrasound is one of the most valuable diagnostic tools available within the clinic today. Undisputedly, it has been adapted to multiple life-sustaining uses that have enhanced medicine and human longevity.

Ultrasound also followed a rather lengthy pathway, not unlike MRI.

The first ultrasound device was built with a piezoelectric crystal by the Curie brothers in 1880. Then in 1916, one of Curie's students built a device to locate a submarine in shallow water. In 1928, Sergei Sokolov demonstrated metal flaw detection with transmission ultrasound measurements and went on to develop and use ultrasound for industrial applications. This device became commercially available in the 1930s.

To produce ultrasound images from the scan, Sokolov developed an imager. This was an array of 1,000 1-mm-wide piezoelectric elements, each electrically connected to a faceplate of a cathode ray tube, which generated an image on the screen of the transmission ultrasound. As a side note, Ferdinand Braun in 1897 developed the cathode ray tube, which was enhanced by Boris Rosing in 1907.

Sokolov's industrial application was not the only use this new technology was put to. As was hinted at by the work of Curie's student, ultrasound technology was under extensive government development beginning in the early twentieth century. In particular, ultrasound came to prominence when work in underwater sound detection of submarines was of interest.

Ultrasound also became an effective tool for detection of flaws in metal structures (e.g., detecting cracks in tank armor plating) and was developed commercially for these and other similar uses.

However, medical applications of ultrasound languished. It was not until 1942 that Karl Dussik adopted Sokolov's technology for medical purposes and built an elaborate apparatus to scan the brain for lesions. To image the upper part of the patient's head, Dussik placed two transducers on either side of the person, who was immersed in a water bath, and recorded the transmission images on photographic paper.

However, the published images did not reveal much information or detail, and the research wound down.

Then around 1949, a U.S. Navy researcher, George Ludwig, received funding to explore the properties of ultrasound in tissue. At about the same time, publicly sponsored research in Japan also attempted to image living tissue and to explore its use in the diagnosis of brain tumors.

By about 1953, there was growing interest in the academic community to research the possibilities of imaging human tissue and organs using ultrasound. All of this was with public and private foundation support, though no industrial interest took shape. This may be in part due to the widely published attempts in brain tumor imaging, which was proving difficult to perform.

The academic work spanned the possible applications including cardiac imaging (University of Pennsylvania, 1957), ophthalmic (University of Illinois, 1956), and National Cancer Institute (NCI)-sponsored research for tumors in 1952.

Concurrently, Ludwig encouraged Ivan Greenwood, an engineer at General Precision Laboratories, to develop a commercial version of a medical ultrasound device.

Ludwig's work in the Navy had caught the interest of John Wild, M.D., a professor at the University of Minnesota. Wild was able to get a military ultrasound device and in 1951 published in the *Lancet* the first images of tissue.

After this early success, Wild met John Reid and drew him into the investigations. Reid was an electrical engineer and saw the possibilities for ultrasound. Notably, his initial work on tissue characterization with ultrasound was done in the 7 years between his B.S. and M.S. degrees.

This two-man interdisciplinary team, funded by the NCI and the U.S. Public Health Service, went on to explore ultrasound application in cancer tissues throughout the body. Reid continued to improve the technology and in May 1953 produced a real-time image of a 7-mm cancer.

Reid and Wild continued as a team until about 1957 when Reid completed his master's. It was a very productive period of work during which Reid not only developed the technology but also expanded its application to cardiology.

Numerous other investigators in the United States—Douglass Howry, Roderick Bliss, Gerald Posakony, and Dr. Joseph Holmes to name a few—experimented with ultrasound imaging of human tissue and organs. A great deal of work was also undertaken in Japan in the 1960s.

However, it was not until Ian Donald used an early device to detect a removable ovarian cyst in 1958 that ultrasound received a great deal of attention. By saving the woman's life, he had clearly demonstrated the clinical relevance of ultrasound.

Soon after, Donald began to use ultrasound for detecting the fetal head, thus launching the first major application of ultrasound in the clinic in 1959.

Like MRI, ultrasound detection was first shown to be feasible in 1916, but 30 years elapsed before any progress toward a clinical device was made. Karl Dussik built an operational ultrasound device in the mid-1940s and published the first ultrasound images in 1947 [108]. Constructing the ultrasound instrument was a multiyear and daunting challenge, and once completed, it was not duplicated elsewhere. It was not until the mid-1950s that ultrasound instruments became more available and could be used by Ian Donald as a diagnostic tool thereby changing the course of history for ultrasound. Perhaps with an early concurrent engineer-clinician team medical ultrasound could have been available decades earlier.

Otto Wichterle: Father of the Soft Contact Lens

What could you build on your kitchen table using spare parts from a bicycle, a phonograph, and your child's Erector Set? You guessed right if you said the soft contact lens.

In 1961, Otto Wichterle developed the basics of a spin-casting system process out of these components, which led to the world's first soft contact lens.

The concept of the contact lens has a lengthy history, packed with many famous players from a variety of fields. From artists, to scientists, to physicians, many searched for an alternative to glasses to correct vision. In fact, many sported glasses themselves. Was it vanity or science that drove them?

It all began in 1508 when Leonardo da Vinci first described and sketched the contact lens. Although he was not interested in correcting vision, the artist is credited for the initial concept.

Da Vinci was followed by the mathematician René Descartes, who suggested setting a lens directly on the eye to correct vision in 1632. Then in 1827, astronomer Sir John Herschel proposed making a mold of the eye that would enable the lens to fit better.

Finally, in 1887, glassblower F.E. Muller produced the first glass contact lens. Soon several others got involved, and in 1936 optometrist William Feinbloom produced a lens that contained both glass and plastic. This led to the development of the first 100% plastic lens, which was made entirely of polymethyl methacrylate (PMMA) in 1948 by optician Kevin Tuohy.

Enter Czech chemist Otto Wichterle.

Born in 1913, Wichterle earned his Ph.D. in chemistry and was teaching at the Technical University in Prague when Germany took control of Eastern Europe at the beginning of World War II. Banned from teaching, Wichterle went to work at the Bata shoe factory in Zlin and began working with polymerization, the chemical process used to combine two or more monomers. It was there that Wichterle invented silon, the Czechoslovakian equivalent to nylon. Unwilling to disclose his invention to the Nazis, as he assumed it would be exploited for use in their war efforts, Wichertle was imprisoned for several months.

After the war, Wichterle returned to academia. Specializing in organic chemistry, he lectured, wrote textbooks, and continued his research. By 1952, he had the idea to use hydrophilic polymers for ophthalmologic use. He recognized that the contact lenses of the time were irritating and uncomfortable due to their lack of oxygen and ion permeability as well as their low resistance to deposition of substances.

He understood that a plastic that comes in permanent contact with living tissue should have high water content, should be permeable, but should not prevent absorption as needed. Working with HEMA (hydrogel polyhydroxyethyl methacrylate), a transparent polymer gel that absorbs water, Wichterle thought he found the answer. Unfortunately, he lacked the mechanism to spin this polymerization into a useful lens.

In 1958, Wichertle once again had his career derailed due to political turmoil in Czechoslovakia. With time on his hands and with the assistance of his physician wife, Wichterle used the materials available to create the spin-casting process by which to manufacture the hydrophilic or "soft" contact lens.

The Czech government sold the rights to Wichterle's invention in 1960 for $330,000. Within 5 years, Bausch and Lomb had purchased the patent for over $3,000,000 and within the decade introduced the world's first soft contact lens.

Today, more than 100 million people worldwide use soft contact lenses to correct their vision. Otto Wichterle received $330 from the Czech government for his ingenuity.

Wichterle died in 1998, but he left behind an impressive legacy. This includes textbooks, a place in the National Inventors Hall of Fame, a school named after him in Ostrava, Czech Republic, and asteroid number 3899 has been named in his honor for his crowning achievements in the field of science.

Harold Hopkins and the Fiberscope: "Let There Be Light"

Admiring that close-up photograph of a lion snoozing comfortably in the sun you snapped while on safari? Appreciative of your regained ability to jog after having arthroscopic surgery performed on your knee?

Pay tribute to the genius of Harold Hopkins, a British physicist and inventor of both the zoom lens and the physics integral to the development of the rod-lens endoscope.

Born the son of a baker in 1918, Harold Hopkins studied physics at University College in England. Although his scholarship was awarded due to his excellence in languages, writing, and history, it was here that Hopkins's talent for science and math prevailed. Graduating just before World War II, Hopkins's skills were put to use developing optical equipment for the British Armed Forces. Subsequent to the war, Hopkins went to work for an optical firm where he developed the first zoom lens for cameras that maintained good picture quality. He went on to teach at the University of Reading in England, where he became the professor of applied optics in 1967 and remained for rest of his professional career.

Endoscopes had been in use since their invention in 1806 by the German physician Philip Bozzini and allowed doctors to see inside the bodies of their patients. They were, however, uncomfortable to the patient and emitted poor image quality causing the medical community to regard them with skepticism.

X-rays, discovered in 1895 by the German physicist Wilhelm Roentgen, were at the time the only practical way to view a patient's bones, organs, and blood vessels. Clearly, another methodology for both improved diagnosis and treatment was essential.

After a dinner party conversation with a gastroenterologist regarding the pitfalls of the rigid gastroscope, Hopkins began working on the concept of flexible fiber optics in 1951. His significant innovation was moving his design from a lens system to a system of solid glass rods with lenses. The rods were bundled glass fibers, each coated with glass of a different refractive index.

This allowed light of any brightness to be guided into body apertures, which improved light transmission while the quality and clarity of images improved significantly. The precision of the images was also greatly enhanced. Additionally, the flexibility of the instrument minimized the invasive and painful impact on patients.

Patented in 1959 in Britain, Hopkins's invention went relatively unnoticed until the 1960s, and Hopkins was unable to obtain financial support for his discovery. In the 1960s, U.S. manufacturers, such as ACMI, and European manufacturers, such as Karl Storz, introduced both flexible and rigid scopes using Hopkins's invention [109]. Since then, these instruments have been widely used for both diagnosis and minimally invasive surgical techniques.

Hopkins's achievement was noted in 1984 when he was awarded the Rumford Medal by The Royal Society. An accomplished pianist, linguist, and sailor, Hopkins died in 1998.

The Accident: Wilson Greatbatch and the Pacemaker

What do Dick Cheney, Don Ho, Phyllis Diller, and Roger Moore all have in common? They, along with thousands of others, have benefited from the genius of Wilson Greatbatch, accidental inventor of the implantable pacemaker.

His life began in typical engineering fashion. Electricity and radios fascinated Greatbatch from an early age, and Thomas Edison was his personal hero. At the age of 16, he even became a licensed ham radio operator.

During World War II, Greatbatch served as a member of the U.S. Naval Reserve, training radio operators and repairing electronics. After the war, he worked as a telephone repairman. Thanks to the GI Bill, Greatbatch earned his B.S. in electrical engineering from Cornell University in 1950. He continued his studies and later received an M.S. in electrical engineering from the University of Buffalo.

While at Cornell, he studied engineering with an emphasis on math, physics, and chemistry. He also began working at the Cornell Psychology Department's animal behavior farm, which is where Greatbatch made the connection between electrical impulses and the human circulatory system.

After his time in Ithaca, Greatbatch returned to his hometown of Buffalo, New York, and became more involved with nuclear physics, battery chemistry, and medical electronics. Working as an assistant professor at the University at Buffalo, Greatbatch also consulted with the nearby Chronic Disease Research Institute.

While using some early silicon transistors to build a little circuit to help record fast heart sounds, he accidentally installed the wrong resistor into the circuit. It started to pulse in a recognizable "lub-dub" rhythm. Greatbatch was already aware of a problem called "heart block," in which the organ's natural electrical impulses don't travel properly through the tissue. He quickly realized that this compact circuit was exactly what was needed to steady these sick hearts.

Before that date, the only method physicians had to regulate a person's heartbeat was a bulky, external pacemaker that plugged into the wall. These antiquated devices also had external electrodes that severely burned a patient's skin.

Using personal funds, Greatbatch refined the device and manufactured 50 implantable pacemakers in his workshop. After 2 years of work, Greatbatch shared his discovery with surgeon William Chardack in 1958. Three weeks later, on May 7, Chardack and Greatbatch successfully implanted their first model in a dog.

Unfortunately, bodily fluids seeped past the electrical tape used to seal the device, shorting it out after only 4 hours. The first internal human pacemaker was implanted in 1960 into a 77-year-old man. He lived for 18 months.

Although Greatbatch credits God for his success, without the earlier invention of the transistor, Greatbatch would never have been able to make his discovery. His interest in the medical applications of engineering gadgets grew from his early days at Cornell University and his affiliation with animal behaviorists.

Greatbatch went on to invent the corrosion-free lithium battery as well as the solar-powered canoe. In 1996, the Lemelson-MIT Program named Wilson Greatbatch as its Lifetime Achievement Award winner.

The inventing of the pacemaker was a highly complicated matter. "Back then I had to solve the problem how to reduce an electronic apparatus in the size of a kitchen cabinet to the size of a baby's hand, in order to be able to implant this pacemaker into the chest," reported Greatbatch.

Philip Drinker: Iron Lung

The year was 1928. The polio epidemic was encroaching upon the nation. Philip Drinker, a chemical engineer from Haverford, Pennsylvania, was employed at Harvard University's School of Public Health.

Specializing in industrial hygiene, Drinker's primary focus was in the area of dust retention in the lungs. Collaborating with a Harvard colleague, Louis Shaw, Drinker designed an iron lung powered by an electrical motor with air pumps from two vacuum cleaners, built at a cost of approximately $500. The pumps altered the pressure within an airtight box, forcing air in and out of the patient's lungs.

The connection to polio was obvious.

Poliomyelitis (polio) paralyzes the victim's respiratory system, usually resulting in death. Unfortunately, their first patient, an 8-year-old girl, died of pneumonia, but only after being successfully revived by placing her in the iron lung.

As described earlier, the iron lung works by placing the patient inside an airtight, metal container; the patient's head remains outside the iron lung, with a seal around the patient's neck. Negative pressure is applied inside the tank to force inhalation and then returns to atmospheric pressure for exhalation. In this manner, the machine mimics breathing and causes air to flow in and out of the lungs.

Previous iterations of artificial respirators included the pulmotor and the negative-pressure ventilator described by John Dalziel, a Scottish physician, in the early 1800s. Dalziel's device used bellows in a sealed box in an attempt to imitate breathing. Alexander Graham Bell (noted inventor of the telephone) developed a hand-operated vacuum jacket, which was never clinically used. In 1876, Julien Woillez devised a hand-operated lung, which was extremely impractical.

These early versions proved ineffective for many other reasons, as well. First, they did not provide enough ventilation and could not be used for an extended period of time. Additionally, they damaged other organs as the air supplies were too powerful.

Working beside a pediatrician, Dr. Charles McKhann, Drinker focused on developing a machine that had the ability to regulate the rate and depth of respiration, was adaptable to many ages and sizes, functioned in a long and steady fashion, and caused no harm to the patient. Taking several weeks to build, Drinker's initial human trials included both Louis Shaw and himself.

By the early 1930s, John Emerson improved on the Drinker Respirator making it both more efficient and less costly. Emerson's version also provided easier access to the patient for both physicians and technicians. Although Drinker, supported by Harvard University, disputed Emerson's patent, he lost his case.

Philip A. Drinker, a biomedical engineer and son of the late Philip Drinker, identifies his father's invention as one of the earliest examples of biomedical engineering, long before the field was recognized. He attributes its success "to the availability of electricity, the immediate need for treatment of polio patients, and the involvement of an engineer in all stages of the device's development."

To say the least, Drinker's ingenuity and design saved the lives of many.

In All . . .

What all of these stories demonstrate is the combination of engineering and medical research in pursuit of tools and technologies to serve in the direct provision of medical care. In some instances, medicine had to wait an extended period of time for the marketplace to produce a commercially viable technology.

In others, mixed teams where science and engineering worked in concert created important medical advances in relatively short periods of time.

While the majority of this book is focused on biological research, these case studies of concurrent engineering in the clinic help demonstrate its potential in the laboratory.

14 Unmet Needs: Mapping and Understanding Cell Signaling

Thus far, the focus of this book has been to examine the history of biological research breakthroughs through the lens of concurrent engineering and then to outline an argument for its potential. We also took a moment to explore the role of concurrent engineering within the context of its contributions and missed opportunities within the clinic.

This chapter is an opportunity to look forward by choosing an example from one of a myriad of biology challenges and looking at how concurrent engineering could be applied. We have chosen the challenge of understanding the basic signaling processes (within and between cells) that govern and control the cell. Clearly, there are many other unmet needs in biology research, so, through this abbreviated journey, the goal is to share an engineering viewpoint on thinking about addressing the needs of biology.

It is not coincidental that this chapter examines a single yet compelling unmet need and then lays out a framework for achieving it via mixed teams of biologists and researchers. This design reflects the suggestion in chapter 12 where we discuss the comparative merits of adopting a DARPA-like model for the NIH.

To wit, we do not advocate a rewriting of NIH's funding methods, but rather incorporation of the specific tendencies that have served DARPA well. In particular, proactively identifying compelling unmet needs, where technology would be put to a novel use, could lead to incredibly valuable results, rather than waiting for such projects to percolate from the research community.

Cell Signaling

In the early 1960s, after the publication of Watson and Crick's structure of DNA, a new era in biology emerged. Researchers began to turn their attention toward divining an understanding of the processes of gene expression and the subsequent development of proteins, which are the building blocks of life. The processes of the central dogma—the expression of genes from DNA to RNA and then into the creation of proteins—became the front and center of biological research.

The past 50 years of advances to discover the structure and nature of DNA and the genes that reside in the chromosomes set the framework for understanding how genes are expressed and how expression is controlled. As outlined earlier, tools were built to obtain the DNA structure.

Aiding this research was the notion that DNA including the genes encoded within it is a remarkably stable structure and can be studied in a nearly static form. For the most part, unraveling the structure and function of DNA did not require dealing with dynamic and rapidly changing variables.

Certainly, over the lifetime of cells and organs, DNA undergoes somatic changes. However, the work of the past 50 years in discovering the nature of DNA did not have to address the transient nature in the genome.

With new questions, though, biologists are seeking answers where dynamic processes are at play. For example, understanding of the production of protein from DNA required a shift in how research was conducted. Researchers moved from the characterization of a nominally static molecule to the characterization of a transient process.

One aspect of this process is the method by which production of a protein is initiated.

The decision by a cell to make a protein from a gene is governed by a highly dynamic process of signals that the cell receives and responds to. The signaling pathways determine critical sequences that control growth, repair, disease, immunity, and so forth. Proteins are not just products of gene expression but also key actors in conveying signals to and between cells and in regulating the expression of genes.

In all, the processes of cell signaling are highly dynamic events that provide a unique range of challenges to a biologist seeking to understand them.

Cells are controlled by signals form the environment as well as by other cells. Signals between cells are

- Juxtacrine—direct contact. Example: "Notch" signaling, which plays a role during organismal growth and development.
- Paracrine—short distance. Example: neurotransmitters.
- Endocrine—systemic or long distance. Examples: hormones such as epinephrine, estrogen, and so forth.

Cells have receptors on the surface and within them, which are activated when they come in contact with a ligand (a binding protein) that targets the receptor.

These can be hormones, neurotransmitters, cytokines, growth factors, and others. The response that an activated receptor triggers can be relatively simple or can be very complex and create a pathway of subsequent dependent signals.

For example, Notch signaling,[1] which is activated by signals transmitted from one cell to another when they are in contact, can activate numerous cell controls depending upon the organ and the type of cell. Another example, which will be looked at again in the next chapter, is the EGF–EGFR pathway, where EGF stands for epidermal growth factor and EGFR is the EGF receptor on the cell. This pathway alone may have a dozen or more steps along the route to regulating gene expression.

The study of cell signaling dates back to at least Martin Rodbell, the 1994 co-recipient of the Nobel Prize in Physiology or Medicine for the "discovery of G-proteins and the role of these proteins in signal transduction in cells." Rodbell began his investigations into the mechanisms of cellular communications or signaling around 1960. Starting with the effects of hormones on metabolism [110], he began to amass an understanding of the different sequences that took place in cellular signaling. The analytic procedure he used was radioisotope marking of metabolic intermediates and signaling molecules of interest.

With Lutz Birnbaumer joining his laboratory in 1967, they began to use chromatographic assays. Through the early 1970s, Rodbell published on the stimulation effects and inhibition of various hormones and then in 1973 on the activation of a guanosine kinase.

Around 1977, Alfred Gilman, co-recipient with Rodbell of the 1994 Nobel Prize in Physiology or Medicine, began his investigations into the details of the signal conduction process. In 1980, Gilman and his team purified the regulatory component of the enzyme adenylate cyclase using ion exchange chromatography [111] and continued exploration using chromatography [112]. This work led to identifying the regulatory function and mechanism of action of this enzyme.

In all, Gilman and Rodbell's work, along with that of many others, served to open the complex study of cell signaling pathways in the late 1970s.

These signaling pathways are at the heart of health and disease. As such, they have become the center of a great deal of research, prompted in part by the hope of identifying signaling pathways that when disrupted can suppress or eliminate disease. Gaining a deeper level of understanding of pathways and their role in disease can lead to developing drugs to control the pathway and thereby treat undesirable conditions in the body.

Because of the potential health care impact, it is little wonder that research in this area has grown tremendously since 1970. This can be seen in figure 14.1:

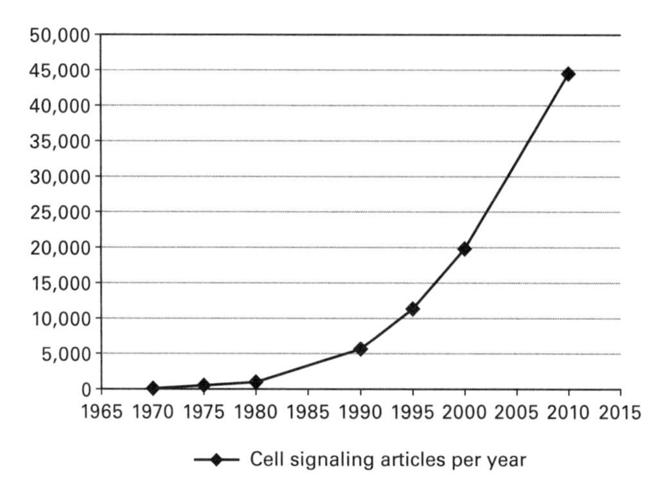

Figure 14.1
Number of articles per year with cell signaling or regulation in the abstract.

The number of articles per year with cell signaling or regulation in the abstract has grown exponentially since 1970.

The Research Challenges

Given the potential impact on health, this book's interest, from a concurrent engineering standpoint, is how scientists can identify and understand the important signaling pathways.

One step in developing a cell pathway map is to identify a gene of interest and then identify the protein that is expressed by the gene. A process of detective work can then be undertaken to identify other proteins in the pathway.

There are several ways in which the expressed protein can be identified. In the case of cancer research, this may start with a mutated gene known to be associated with cancer. This can be found by bioinformatics analyses of sequenced genomes of both healthy and sick patients by comparing and finding differences in the genome that correlate with the disease. Currently, numerous research facilities are working on creating just such an atlas.

Once the mutated gene of interest is identified, scientists can simulate the likely protein that would come from this gene. This, however, is fraught with difficulty. There are considerable alterations of the transcribed gene that take place in subsequent steps such as RNA splicing and editing. Moreover, epigenetic alterations of the DNA message can affect the expression of the gene and hence alter the levels of the protein product of this gene. Importantly, after translation, proteins can be decorated with tags (modifications), which can alter both their stability and activity. Therefore, finding how a mutation affects a protein function can be a significant and complex undertaking.

Another approach is to create an animal model with a switchable mutated gene and observe the proteins that are created or eliminated. The expression of the normal gene is suppressed through genetic engineering and replaced with the mutant version. By comparing the protein function in the modified animal versus that in a control animal, one can infer the effect of the mutation. This, however, is also a challenge as animal models do not necessarily recapitulate human biology.

The study of cell signaling is further complicated by the fact that it involves a variety of molecules along the pathway. For example, many pathways have been found to include a protein activation step in which a molecule (often a kinase) will alter the folding of a protein to expose or hide the protein's active site. Thus the kinase can turn on or off the active state of the protein, which it does by adding a phosphate group to a protein, thus phosphorylation.

Phosphorylation is one of many elements of signaling pathways but is the focus of a good deal of research and is a good case study to understand the challenges of monitoring cell signaling.

Monitoring Cell Signaling and Phosphorylation of Proteins, 1954

In 1954, George Burnett and Eugene Kennedy showed [113] that there is increased activity of proteins after phosphorylation using radioactive labeling with phosphorus 32 (^{32}P). This seemingly isolated but very important discovery became the focus of research because phosphorylation is one of the basic and major process steps in cell signaling and gene regulation and is at the heart of many pathways.

The addition of a phosphate molecule to an amino acid can turn a hydrophobic portion of a protein into a polar and extremely hydrophilic portion of the molecule, thus introducing a conformational change in its structure. This step can promote or inhibit the protein's interaction with other molecules.

Furthermore, they found that a protein kinase is capable of moving high-energy phosphate groups from adenosine triphosphate (ATP) to a substrate, which can then power subsequent functions.

In all, these findings were the dawn of the research of the past 30 years to identify the pathways that actually regulate the expression of genes and subsequent downstream proteins.

Measurement of Phosphorylation via Western Blotting

There are several ways to measure phosphorylation. Probably the most commonly used method is the so-called Western blot, which is an advancement of gel electrophoresis technology developed in the 1930s.

It is also one of several specialized gel electrophoresis techniques (table 14.1). The first technique was developed by Edwin Southern to separate and identify DNA fragments. The initial Southern blots were done with radioactive-labeled DNA fragments [114]. Subsequently, Western blot was developed for protein as opposed to DNA analysis.

In Western blotting, proteins are first arrayed using an electrophoretic gel, and then a protein is recognized using an antibody, specific to the protein of interest. Finally, the antibody is detected using a fluorescence label or by a chemical reaction. Phospho-specific antibodies have been developed in order to detect the phosphorylation state of proteins. Once again, the target protein as well as the specific amino acid that is phosphorylated must be known in advance so that the correct antibody label can be applied.

Figure 14.2 shows a simplified version of the general concept of a Western blot.

Another approach is the enzyme-linked immunosorbent assay (ELISA), in which a lysed sample is used and sandwich assayed with phosphor-specific antibody and a fluorescent marker.

Table 14.1
Specialized gel electrophoresis techniques

Technique	Separation	Identification
Southern blot	DNA fragment (from restriction enzyme)	Complementary DNA + label
Western blot	Protein ("immunoblot") and SDS-PAGE gel	Antibody + label
Northern blot	RNA	Complementary RNA or DNA + label

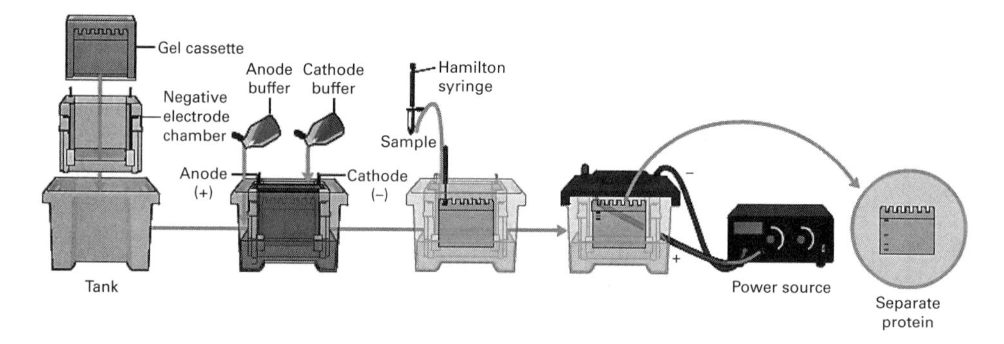

Figure 14.2
Typical Western blot setup.
Source: License under Creative Commons. Creator: Bensaccount at en.wikipedia.org.

Then there is flow cytometry, in which fluorescent markers, such as those bound to antibodies specific to the phosphorylated entities on cells, can be identified using light-activated gated cell sorting.

The latest development in signaling pathway technology is the use of mass spectrometry to detect the presence of molecules in a cell as a result of some stimulus. Mass spectrometry has the advantage of being "agnostic," meaning that you do not have to identify the protein you are looking for in advance, as you must with Western blotting. However, this can be an extraordinarily complex undertaking as the sample size necessary for mass spectrometry is many millions of cells, which can be heterogeneous and thus express a mix of proteins.

Each of these techniques has a specific advantage and application, and several may be used in any particular investigation.

An example of a research project on a pathway will be instructive so the reader can develop a sense of the sheer complexity and enormity of the effort to understand the regulatory pathways.

Example: Tumor Necrosis Factor and Apoptosis Pathway Study

This example was taken from an MIT research team composed of Kevin Janes, John Albeck, Emily Pace, Doug Lauffenburger, Peter Sorger, Suzanne Gaudet, and Michael Yaffe. The work was a study of the signaling pathway triggering cell death from a stimulus of tumor necrosis factor (TNF), insulin, or EGF.

These factors were known to affect the cell death cycle, but their role in the regulatory process was not well understood at the time of the research. Based on prior work, the team decided that approximately 19 different signaling nodes should be examined at 13 different time points and this done in triplicate. Because cell signaling is a dynamic process, it is important to make multiple measurements over time as the individual activations may change. Taking a snapshot of a single time point for a dynamic process can be very misleading, as illustrated in figure 14.3. Moreover, the three stimuli—TNF, insulin, and EGF—needed to be taken in pairwise combinations in order to examine their additive effects, thus resulting in multiple stimulus experiments.

Altogether, some 7,000 experiments needed to be conducted in order to characterize the impact of each of the three stimuli on the apoptosis (process of cell death) of the cell.

The 19 targets were largely kinase-activated targets, and the approach used to measure the activation was Western blotting. A 96-well format for the gel was developed to prepare the samples for Western blots in order to accommodate the myriad measurements that had to be made. Additionally, some 700 Western blot arrays had to be prepared and analyzed. A typical Western blot

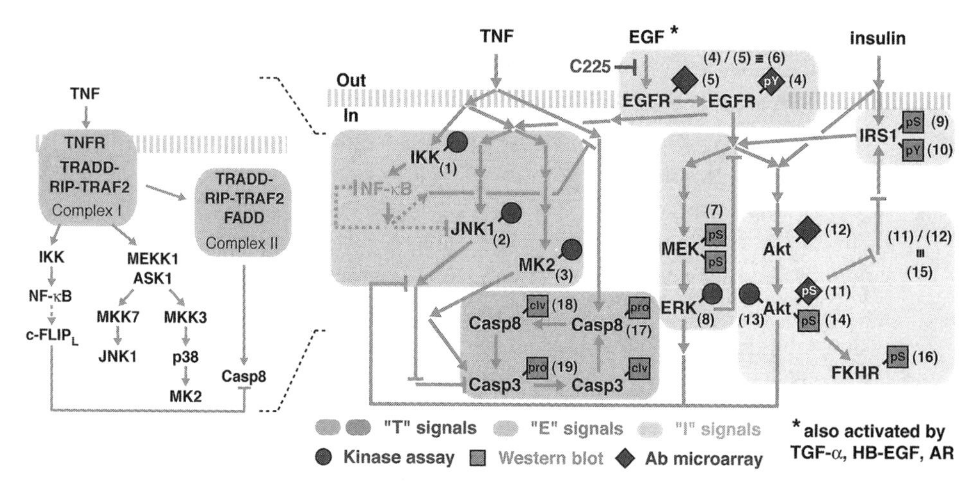

Figure 14.3
TNF cell signaling pathway and the types of assays that were used to characterize the pathway. Nearly 7,000 Western blots were done as well as numerous other assays needed to develop the pathway information.
Source: From Kevin A. Janes et al., "The Response of Human Epithelial Cells to TNF Involves an Inducible Autocrine Cascade," *Cell*, March 24, 2006, p. 1226. Reprinted with permission of Elsevier. Copyright © 2006 Elsevier.

requires about 1 day of preparation and 1 day of analysis. Even doing many gels in parallel would consume hundreds of person-days.

As shown in the subsequent papers by this team, one could reach any one of three conclusions about the effect of TNF on cell death:

- There is no effect of TNF on cell death.
- TNF correlates positively with cell death.
- TNF correlates negatively with cell death.

The range of conclusions is due to the fact that at different times, TNF alters the signaling process differently. This is a profoundly important observation, pointing to the time-varying nature of cell signaling. Nearly all prior breakthroughs, such as the sequencing of the human genome, were based on measurements of a time-invariant biological element.

With cell signaling, it is necessary to look dynamically at the interaction of numerous pathways and make numerous parallel measurements over time. The current technologies for cell signaling measurements are end point in nature and therefore require repeated labor-intensive experiments.

What this points to is a significant unmet engineering need in biology—a technology that can identify multiple proteins and assay for their activation state and function in real time as the process unfolds.

A New Approach Is Required

Michael Stratton of the Sanger Institute estimates that there are approximately 40 mutated driver genes that are important in breast cancer alone that we know of at present [115]. Assuming that it requires 100 person-years to characterize the pathway for each of these, including the interactions among the pathways, we are looking at a gargantuan effort in order to understand the signaling pathways in breast cancer based on the current technologies.

Obviously, this is an incredibly important task for engineering.

In Vivo Cell Signaling—Real Time

To an engineer, looking at the complexity of the signaling network, it would seem that a technology capable of identifying dozens or even hundreds of signaling events in real time and in vivo (initially in animal studies) would be an important goal. You may say, "Well, of course that would be wonderful, but is it possible in our lifetime?"

You could have said the same thing in 1960 with regard to landing a man on the moon and returning him safely home.

While advances in mass spectrometry may offer a much better approach to pathway discovery than Western blotting, it will still be a long, slow process. However, as we have discussed earlier, there is no telling what novel technology exists or is being developed within another domain of science but has not yet crossed over to biology. There is also the possibility that a promising technology, again in another domain of science, will never see the light of day because it is perceived to lack commercial viability.

It is not too incredible to think that perhaps there is a novel device in the basement of a physics or electronics laboratory somewhere in the world that is not thought to be relevant to biology. However, if applied, it could allow biologists a completely new comprehensive view of the dynamics of pathways in real time.

Our study suggests[2] that if such a technology existed today, then left to the current processes it might be 40 years before it is available to a biologist to make the breakthrough we aspire to. Forty years unless we embark on concurrent engineering to draw out this new technology and bring it to the biology laboratory in 10 years.

There is no current funding approach to bring this about. Similar to the genome project, this will require a DARPA-like directed program. Such a program would have an objective laid out with phased funding to support 10 to 30 seedlings. Those that succeed would be culled out and go through a phased process with clear, objective milestones to narrow the projects to the finalist.

Systems Biology

Systems biology can help us understand the major pathways that are important and should be the target of our efforts, thereby streamlining the search for the important regulators.

In a study cited by Doug Lauffenburger [116], head of MIT's Department of Biological Engineering, systems biology techniques were used in research on three kinds of lymphomas. The researchers found that 80% of the roughly 65,000 network interactions appeared to be unaffected by any of the tumor phenotypes. Meanwhile, hundreds of interactions were differentially correlated. Systems biology offers the potential of rapidly narrowing the near-infinite number of protein interactions that could be important in any particular disease and thus accelerating the research and discovery process.

Lauffenburger's findings suggest that systems biology is a powerful tool for narrowing down the search for effective targets and for understanding the emergence of alternative pathways that produce treatment resistance.

Systems biology offers the possibility of reducing the timeline to more effective treatment. However, the problem of cell signaling is far more complex than the systems engineering challenges of the manned space program, which helped define the scope of systems engineering. The biological systems problem is the reverse of the NASA problem. At NASA, humans designed and understood each layer, and systems engineering allowed them to understand the aggregate function. By contrast, biology works with systems components beyond our current understanding, thus biology researchers have less working knowledge of each piece of the system.

The German physiologist and Nobel laureate Otto Warburg observed that unlike normal cells, cancer cells use anaerobic metabolism even in the presence of oxygen. In a 2009 paper, Matthew Vander Heiden, Lewis Cantley, and Paul Thompson [117] discuss the implications of the Warburg effect in cancer. They speculate that a small number of metabolic pathways might govern proliferation of many cancers and offer a less manpower-intensive way to address cancer therapy than silencing the numerous phosphorylation pathways. Perhaps connecting systems biology and characterization of cell metabolic processes could yield a robust view of important pathways more quickly.

As discussed in a case study in the next chapter, kinase (recall that kinases are proteins/enzymes that activate other proteins by phosphorylation) inhibitors have been approved and marketed for use in cancer treatment and are widely regarded as an incredibly successful drug development.

Clearly, a new approach is needed. A real-time signaling pathway analytic technology guided by systems biology could dramatically reduce the discovery time. Ideally, the technology would be applicable to in vivo analysis. Further complicating this approach, the technology would need to provide real-time cell signaling information on a cell by cell basis for thousands, perhaps millions of individual cells in the normal and tumor microenvironments.

What is undoubtedly needed at this point in time is an integrated effort of engineers and biologists. These teams would model the likely pathways and the manpower effort needed to characterize them as well as develop new technologies to reduce dramatically the characterization effort.

Systems Biology—A Model to Guide Research

In addition to the challenge of measuring the signaling pathways, there needs to be effective analysis of the mountains of data, which is one role that systems biology can play. The author interviewed Dr. Linda Griffith about the use of systems biology to guide her research:

> The Alliance for Cell Signaling that Al Gilman was in charge of, the idea was to make all these signaling measurements, these huge numbers of measurements. Then as Al Gilman put it, I was on a panel with him, you turn the crank of calculus and the answer comes out, because you have all the measurements.
>
> However, that particular [NIH] Glue Grant [to the Alliance] hasn't had a lot of impact because you have to think about really, there wasn't the framework for thinking about how to interpret. It's not just turning the crank and calculating—it's not just having a mathematician having a really efficient computer program—it's really thinking [at] a much higher level of abstraction.
>
> It's having a model of what is going on and what guides you in how you set up your experiment. It guides in what kind of cell you would pick. It tells you what your cue space should be. It tells you all of these things that help make your system that you're doing your measurements on generalizable to a lot of other systems.
>
> So that's part of the issue. In biology, it is complex, so there is a tendency to think something is idiosyncratic or particular. What the engineers bring in is to think about what can you generalize about. Of course there are different things, you wouldn't have all these different cells if there weren't some specific things, but the specification comes about on a chassis, if you will. You built the motorcycle and it's your idiosyncratic special motorcycle, but you started with some kind of basic construction parts and added on to that and it's sort of the same way you can think about cells. How do you describe the operating system without the bells and whistles that make it a retinal cell?
>
> So I think where it's helpful is to parse what shouldn't really be a nuance. There's a difference between providing biologists tools and engineers making tools . . . [there is value in] working as a team to understand how to cast the problems originally when we're going to generate data. How do you do the measurements so that you get to where you can start to build real models?

When asked if this kind of collaboration is happening, Linda says:

It is happening . . . and again there is always inefficiency about how things work. NCI, for example, is funding a lot of these [programs] putting physicists together with biologists. There is unfortunately not an appreciation at NIH that physicists are also pretty reductionists. So you've got a bunch of reductionist people, and engineers by nature try to look at complicated problems.

There is still, I think, not quite the appreciation at NIH that [biological system] engineers like ours exist and are there to build a framework as opposed to building the tools. NIH feels like if there's any kind of math involved, meaning an abstraction or theory, it must be a physicist who will do it or a mathematician, and not quite percolating in *that it's actually the engineers that build the kind of models that describe the system.*

Physicists are sort of in the same category as biologists in terms of the way they parse complex problems. I'll put it that way, it seems NIH still thinks of engineers as people who build things. I wrote a review in *Science* ten years ago where in engineering you need the design principles. You can't build things without them. Maybe by luck you'll get something to work. However, if you really want by luck you could have landed on the moon without the systems engineers, but you'd have to be pretty lucky, right?

So it's really that systems, that kind of parsing the problem and getting any of this biology to work requires mechanical engineers who know how to make a particular machine, or an electrical engineer. So the biological engineer alone communicates with the biologists and communicates with the other kinds of engineers that need to contribute to the whole thing. And it's the paradigm that it's a classic engineering paradigm and then you have a revolution of the underlying science.

Mission to the Moon and the War on Cancer—Systems Engineering

Systems engineering grew out of the complex developments, undertaken during the second half of the twentieth century, principally around supportive technology for World War II. Systems engineering was involved in an even greater number of activities as the space program came into being in the mid-1960s. It was recognized that numerous engineered components would interact in complex ways and needed to have high reliability and predictability. This was most dramatically seen in the Apollo program to land a man on the moon.

With millions of important interactive components in the Apollo spacecraft, no one could comprehend or predict the outcome of any one particular event without some type of model that simplified the interactions so that it could be grasped by an individual. By abstracting the component interactions to a high enough level, a context for understanding outcomes and interactions that an individual could grasp could be created. Then when a malfunction occurred, the systems engineer could take down several layers of complexity to a specific subset and determine what failure could have led to the observed failure he was investigating.

President Kennedy announced the mission to the moon in 1961, and Neil Armstrong walked on the moon in 1969. Thereafter, President Nixon announced the mission to cure cancer in 1973, and we have yet to achieve this goal. This is not to suggest that both problems are equally complex, but to address the problem it is interesting to look at the governance structure for each program.

The Apollo program has been lauded as a tour de force in systems engineering and systems management. A framework was created at the very outset that allowed for 100,000 scientists and engineers to collaborate on complex systems to address the challenge. To say the least, there were many scientific and technical unknowns at the outset in 1961.

To address these unknowns, there was simultaneous research in the areas of human performance in zero gravity, materials behavior on the lunar surface, and propulsion systems appropriate for the lunar surface, to name a few. Of course, there was also the close interaction between the various experiments and research within the engineering community that needed to build the technology and deliver it to the lunar surface.

This is not to say that curing cancer is an engineering problem. However, perhaps organizing the considerable resources conservatively estimated at nearly 150,000 researchers and technicians in 11 states working on solving cancer would benefit by a systems engineering framework. This would allow individual researchers to collaborate with engineers more readily as the context of the problem would be more easily shared between these diverse disciplines.

15 Unmet Needs: Cancer Example

In the previous chapter, we took a slice through biology with a thin knife by thinking about characterizing cell signaling as a major research challenge. In this chapter, we take another cut through biology with the challenge of cancer in mind.

Cancer is widely regarded as a disease of the genes because genetic mutations are necessary for the proliferation and growth of cancer. Obviously, this genetic connection has placed a significant focus on the human genome as well as the continually improving sequencing technologies. These sequencing technologies not only are far more effective, but also they cost less and are leading to ever greater understanding of the genes involved in the broad array of cancers.

The expectation by some is that the ability to map the genome inexpensively will mark a turning point in our war on cancer. However, there is a significant question as to whether it will deliver clinical results without the development of other new technologies (beyond sequencing) from the engineering side of the house such as improved means for observing cell signaling and other processes that regulate gene expression into proteins.

To understand the unmet needs for enabling technology in cancer research let's look at the history of cancer research, clinical advances, and the technology that supported the progress so that we might have some context for the challenges that the author believes that engineering can address in the fight against cancer.

While there are numerous Nobel Prize–winning breakthroughs in genetics, there are very few in cancer. Still there is much to be learned, and so an alternate source for cancer milestones was sought and found. The journal *Nature* has published a review article [118] on the milestones in cancer research and treatment (see http://www.nature.com/milestones/milecancer/index.html for a collection of materials), which will be our road map to explore whether and how engineering has enabled cancer research and therapy.

Finally, unlike the study of genetics, where the end goal is improved knowledge of the function and structure of genes, cancer is a disease, and we must understand not only the biology but also how to control or cure it.

Milestones in Cancer, 1893 to Now

The first major discovery milestone in cancer typically traces to Peyton Rous and his discovery of tumor-inducing viruses in the early years of the twentieth century. For this work, he was awarded the Nobel Prize in Physiology or Medicine in 1966.

However, there was an earlier discovery in 1893 by Dr. William Coley, a young surgeon at New York Memorial Hospital [119]. This story is interesting for several reasons. Coley's remarkable discovery led to a cancer therapy that continues to be used to this day even though the mechanism of action is not completely known. Several hypotheses have been suggested, but still no mechanism has been identified.

Coley was treating a young woman with a sarcoma on the arm. After surgical removal and no outward signs of cancer, the disease still progressed rapidly, and she died within 2 months. Coley was upset by the outcome and decided to research the medical records of similar patients.

Much by chance, he found the case of a man with a large sarcoma, which was resected, reoccurred in a different location, but whose wound from surgery became infected. As the infection grew worse, the other tumor was observed to regress and then disappear. Four months later the patient was discharged from the hospital with no infection and no cancer.

Coley suspected that the infection drove the cancer into regression. He experimented with patients by infecting them with bacteria and observed that the tumors often shrank, though the infections were sometimes fatal. To develop a safer remedy, he sought and found a mixture of two killed bacteria that would induce the inflammation and immune response without inducing the virulence.

The Coley vaccine as it became known was used successfully to treat carcinomas, lymphomas, and melanomas. The vaccine was injected directly into the tumor or metastases over a 6-month course of treatment. With continued work in the area, it was found that Bacillus Calmette-Guérin (BCG), the now widely used tuberculosis vaccine, was an effective treatment for early bladder cancer.

Now onto the more famous work of Peyton Rous.

In 1911, Rous was working with sarcoma in chickens when he discovered that he could initiate sarcoma in healthy chickens by transferring filtrate from homogenized excised sarcoma tissue. The filtrate obviously contained particles smaller than cells, and Rous hypothesized that the cancer-producing element that was transferred must be a bacteria or a virus.

His work was vitally important to our modern understanding of cancer as a genetic disease. It also established the foundation upon which subsequent work was based, such as that by Harold E. Varmus and J. Michael Bishop. Varmus and Bishop received the 1989 Nobel Prize in Physiology or Medicine for their discovery of the cellular origin of retroviral oncogenes.

1900–1915

The next milestone in cancer discovery was the work of Abbie Lathrop and Leo Loeb in 1916 [120]. These two researchers explored the origins of tumors in mice, which led to discovery of the effect of hormones on cancer.

Hormones are molecules secreted by cells to signal changes in other cells. They can be proteins (transcribed from genes) or lipids (created via chemical synthesis). Later, it was determined that the secretion that they observed as the cause of cancer was testosterone in men and estrogen in women. They compared castrated mice with noncastrated mice and observed that internal hormonal secretion is a factor in the incidence of cancer.

1960

Surprisingly, the *Nature* milestone timeline does not identify another milestone experiment until 1960. This is when Peter Nowell and David Hungerford [121] reported the discovery of a chromosomal aberration, which became known as the Philadelphia chromosome.

The importance of this discovery is well documented by a brief history published by the University of Pennsylvania Health System [122]. In essence, the aberrant chromosome was shown to be present in the cells of leukemia patients but not in normal cells. Nowell and Hungerford's careful study demonstrated that cancers are caused by a gross alteration of genetic structure, which by the way is not inconsistent with the previous observation by Rous that the virus causes cancer as it embeds genetic material capable of inducing transformation into the host DNA.

1970

The groundbreaking work on the Philadelphia chromosome set the stage for an explosion in the genetic origins of cancer. Through the late 1960s and into the 1970s, numerous researchers identified key genetic causes to cancer.

In 1970, G. Steven Martin at the University of California at Berkeley published [123] a paper identifying the probability that the Rous sarcoma virus contained a gene that gave it tumorigenic properties in chickens. This type of gene became known as an oncogene.

Around the same time, across the San Francisco Bay from Martin, J. Michael Bishop and Harold E. Varmus at the University of California at San Francisco were investigating the characteristics of the Rous sarcoma virus. In 1976, they published [124, 125] the first two of their papers, which identified the gene in the Rous sarcoma virus responsible for the cancerous behavior.

They found the oncogene in the Rous sarcoma using hydroxyapatite chromatography. This technology is similar to chromatography but uses hydroxyapatite resins (introduced in 1956) to enhance the separation of viral DNA. Although chromatography was first demonstrated in 1900 by Mikhail Tsvet, the hydroxyapatite resins were first demonstrated in 1956. Then in 1978, Bishop and his team [126] identified a nucleotide sequence in the sarcoma virus that confers the oncogenic capability.

What these stories on cancer show is that unlike the history of molecular biology, the war on cancer has not benefited from much concurrent engineering aimed at addressing the problems sometimes unique to cancer. So, how can engineering enable our understanding and treatment of cancer ?

How Engineering Can Enable Cancer Biology?

The challenges in addressing cancer can be thought of in the broad categories of early detection, characterization (genetic and phenotype), and (targeted) treatment.

Early Detection

Early detection is often cited as an unmet need in cancer research and treatment, but just how early is early?

Among his many important discoveries, Dr. Judah Folkman [127] showed that for a tumor to develop to a malignant and deadly phenotype, it must first recruit its own blood supply. Quoting Dr. Folkman: "Most strikingly, microscopic carcinoma (many of them less than 1 mm in diameter) is found in the thyroid of more than 98 percent of individuals age 50 to 70 years who die of trauma, but is diagnosed in only 0.1 percent during life in individuals in this age range."

Through examination of numerous tumors, Folkman concluded that tumors less than 1 to 2 mm in size lack blood vessel support and remain dormant until they recruit blood vessels. One would in principle want to detect tumors before they recruit blood vessels and send cells (metastasis) throughout the body. However, detection of a tumor 1 to 2 mm in size is an unmet need.

Early detection of cancer is important for several reasons. Prior to metastasis, the homogeneity of the early population of cancer cells lends itself more easily to treatment. As the primary tumor cells proliferate, they acquire variations in the genotype due to the instability of the genomes of cancer cells, and a heterogeneous population is created. Because drugs are typically directed to act on one or two mechanisms of the cell's processes, the heterogeneity of tumors represents a huge challenge in the clinic. Heterogeneous tumors are more likely to acquire drug resistance and regenerate the cancer as a reoccurrence later. Early detection can improve the likelihood of a successful outcome as it would lead to treatment at a time when drug resistance is minimal and reoccurrence is less likely.

As Folkman [127] points out, tumors that reach the 1- to 2-mm size range begin to acquire blood supply and the capability to grow and have access to blood vessels to metastasize. Therefore, the question is whether we can detect 1-mm tumors.

Imaging

Today's PET-MRI scan resolution is greater than 10 mm [128], which indicates that *a 10-fold improvement in resolution is an unmet need.*

Sanjiv Gambhir [128], director of Molecular Imaging at Stanford School of Medicine, was able to detect tumor lesions in a transgenic mouse as small as 3 mm in diameter. To do this, he used a laboratory imager with a 1.6-mm spatial resolution and an experimental imaging agent, Cu-DOTA-knottin 2.5F.

This level of imaging approaches the desired 1-mm level for early-stage detection, but the cost of PET-MRI would be prohibitive for healthy population screening.

There are hyperpolarized agents that can increase MRI sensitivity by 10,000-fold [129] and offer the possibility of early-stage detection and the 1-mm level of resolution. However, there appears to be no concerted effort to make it a clinical technology.

To detect early (1-mm) tumors in a cost-effective clinical setting by imaging would appear to be beyond the current technology and probably requires a paradigm change in how we think about cancer detection.

Gene Expression (mRNA Detection)

Detection of gene expression holds the promise of being the basis of a cancer detection technology.

The mRNA is contained in cancer cells, and therefore this approach requires access to the tumor cells. This is feasible in certain cancers that shed cells in fairly large numbers via exhaled breath, urine, or feces. Allegra Diagnostics (http://www.allegrodiagnostics.com/), for instance, is able to detect lung cancer from cells in the airway passages.

This method is not feasible for breast, brain, skin (because it must be localized), pancreatic, uterine, and other cancers because the shed cells from the primary tumor are not easily captured.

Biomarkers

In an attempt to find low-cost ways to detect cancer at an early stage, a great deal of work has gone into identifying biomarkers such as proteins, mRNA, or DNA fragments.

There are generally two classes of biomarkers: molecular markers (such as proteins, mRNA, or DNA fragments) and cellular markers. In a study published in 2008, Gambhir [130] demonstrated in a mathematical model that, based on current detection limits of proteins, our "early" detection limit using blood-borne biomarkers would be a tumor about 3 to 70 mm on a side (27 to 340,000 mm^3). This is discouragingly far from our 1-mm goal.

With improved sensitivity (0.01 U/mL for CA125 vs. current immunoassays averaging 0.7 U/mL) and with highly specific biomarkers, a 0.5-mm tumor could be detectable. However in a more realistic clinical scenario, the detectable tumor is about 5 to 10 mm. This is about 1 billion cells, which if starting from a single cancer cell would mean about 30 doubling events (doubling number = $\ln 10^9/\ln 2$). The typical breast cancer cell doubling time is 50 days, so the cancer has grown undetected for 1,500 days, or 4 years and would not represent a very early detection. Leading-edge nanotechnology research might yield a means of amplifying the biomarkers and overcoming the dilution problems that Gambhir points out.

Butler [131] found that in breast cancer, tumors shed about 3.2×10^6 cells per day per gram of tissue and a 10-mm tumor (1 cm on a side, or 1 cm^3) weighs about 1 g, so an average of 3 million cells per day are being shed into the blood by a mere 1-cm tumor. This provides a lot of opportunities for spreading into other organs prior to primary tumor detection.

On the other hand, could this massive number of shed cells be used to detect the early onset of cancer? These circulating tumor cells are the focus of a number of studies and the next subsection.

Circulating Tumor Cells for Detection and Characterization

While a 10-mm tumor may theoretically shed 3.2×10^6 cells per day, the cells probably only survive in the blood for one circulation through the body, or about 5 minutes. So at any one time, there might be 1.2×10^4 cells in circulation.

A typical blood draw is 10 mL (there is 5 L of blood in the body), so theoretically we might see 24 circulating tumor cells (CTCs) in 10 mL of blood from a 10-mm tumor. However, 10 mL of blood contains 10 billion cells, so the challenge is to find the 24 CTCs out of 10 billion! This is a monumental challenge, and it must be done very cheaply (say $10 per test) if it is to be a regular cancer screening test.

This challenge is being approached by a number of programs. Johnson & Johnson already offers a commercial instrument for identifying CTCs. Meanwhile, Dr. Mehmet Toner [132] at Massachusetts General Hospital has demonstrated a microfluidic chip technology for isolating CTCs.

In a 2011 paper [133], Toner and his research partner, Dr. Daniel Haber, speculated: "As the technology evolves, applications in the early detection of invasive, but localized cancer may even be contemplated."

This opens the possibility that CTCs can be used for early detection of invasive cancer.

The question is at what stage does the primary tumor begin to shed cells? As Haber and Toner point out, the number of circulating tumor cells they measured does not correlate well with tumor mass and may reflect tumor vascularity or invasiveness. This suggests that a small undetectable (with current imaging) primary tumor may be vascularized and shedding tumor cells.

A limited amount of modeling of primary and secondary growth rates suggests that to explain the size of secondary tumors, the dissemination of cells must have begun when the primary tumor was 1 to 4 mm [134]. This is consistent with the Folkman work showing vascularization with tumors of this size. In this model of *parallel progression*, tumor cells are disseminated when the tumor diameter is 1 to 4 mm, and seeding takes place in different organs with some developing within 6 years. This model further suggests that *factors secreted* by the primary tumor may stimulate growth.

In their earlier paper, Haber and Toner [135] showed a mean tumor burden (size of tumor) of 8 cm diameter and 90 CTCs/mL for 21 patients. This is a very large tumor burden. Assuming that shed rates and CTC numbers scale with the tumor mass, we would expect about 1/1000 of that amount from an early-stage 8-mm tumor, or about 1 CTC per 10 mL. That would be very challenging to measure, but it may be what is needed for early detection.

CTCs also offer the researcher a much needed population of tumor cells for studying and understanding the mechanics of metastasis. These cells, if recovered, can be studied and used to guide drug development and therapy strategies to control or eliminate metastatic cancer. The engineering challenge is to develop an instrument that can find 1 cell with metastatic potential out of 10 billion blood cells.

Characterization

Characterizing the genetic and phenotypic nature of different populations of cancer cells is vitally important to gain an understanding of the biology of cancer and how to treat it. As such, it represents another important unmet need where concurrent engineering could play a valuable role.

Image tumor cells in vivo continuously and at the single-cell level Small subcentimeter confocal microscopes [136] offer the possibility of continuous monitoring of biological events in vivo and for extended periods of time. These miniature microscopes could detect molecular markers on a single-cell level while maintaining a wide field of view of hundreds of cells. The instrument can be implanted in the laboratory animal and the cellular events observed over an extended period of time. This technology can reveal much about the propagation and growth of cancer.

Methods to activate functions in vivo remotely Scientists wish to be able to turn on and turn off gene expression in animal models with an external control (such as a specific wavelength of light) in order to study specific pathways and drug efficacy.

Model systems The mouse has been widely used for decades as a model system and continues to be a major source of discovery and drug development. However, the mouse model does not always recapitulate the human response because "80 million years of independent evolution separate us from our rodent cousins" [137]. We need ways to characterize cancer and test hypotheses with human organs and/or tissue or in humans. Perhaps microsystems, including microdosing, can be used to understand cancer development and control.

Measurement of the metabolic rate of a single cell and group of cells in a tumor Cancer cells are known to have higher rates of metabolism (glycolysis), commonly referred to as the Warburg effect. Otto Warburg proposed the concept, and recently there has been some elucidation of the mechanisms. Technology is needed that will measure the metabolic rates both for detection and characterization.

Characterizing the regulatory pathways of cancer As Robert Weinberg and Douglas Hanahan show (discussed in the next section), these regulatory interactions are similar to the circuitry of complex integrated circuits. This suggests that, at some point in the future, biological systems engineers may help unravel and understand the complexity of these regulatory or cell signaling systems.

Drug Development and Delivery

It was not until the development of our understanding of regulatory pathways in cancer (circa 1970) that we could develop targeted therapies for controlling cancer. Genotoxic chemotherapies were used to attack rapidly growing cancer cells preferentially, while leaving many of the normal, more slowly growing cells unharmed. Still, these therapies have significant and sometimes fatal side effects. Additionally, they have proved to be ineffective in cases of reoccurrence or metastatic spread of most solid tumors.

In their discussion of the hallmarks of cancer Weinberg and Hanahan outlined the dynamic regulation system that initiates the expression of the genes of interest. Figure 15.1 gives us some idea about some of the regulatory pathways (at a very high level) discussed in the Weinberg and Hanahan paper.

In this figure, we see the role of several prominent proteins (Ras, Myc, p53) in the key processes of cancer. Mutated genes of these proteins produce unwanted behavior and are under the control of regulatory processes. Signals from surrounding cells and stroma affect the expression of these genes and the production of proteins, which cause the proliferation of additional cancer cells

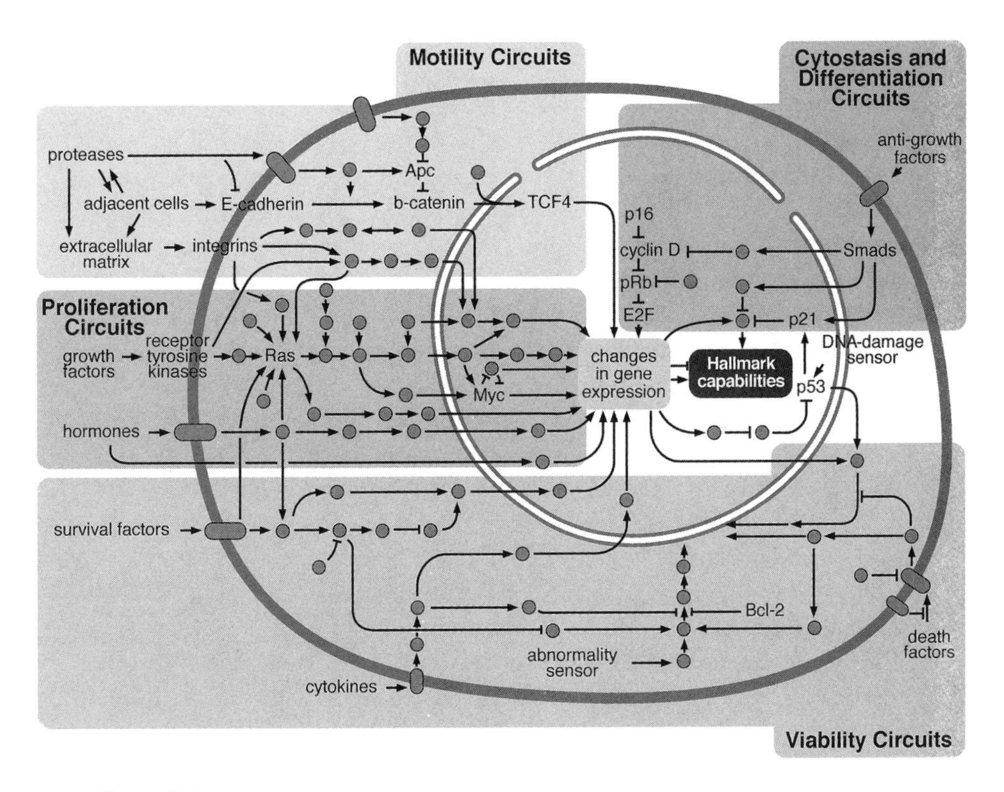

Figure 15.1
Intracellular signaling networks regulation the operations of the cancer cell.
Source: From Douglas Hanahan and Robert Weinberg, "Hallmarks of Cancer: The Next Generation," *Cell* 2011; 144(5):646–674. Reprinted with permission from Elsevier. Copyright © 2011 Elsevier.

and their dissemination. Shown in the regulatory route are examples of important pathways for the various hallmarks of cancer.

This represents a small fraction of the known tumorigenic pathways. Thus, the reader can appreciate the growing complexity of our understanding of cell regulatory pathways, particularly as it applies to managing cancer. The diagram shows some 60 key protein steps (little circles) making up these regulatory pathways.

Weinberg and Hanahan identified approximately 10 therapeutic targeting approaches for cancer, and we will touch on a few here in order to develop a sense of the challenges.

Immune System Therapy—Stimulatory

The Coley vaccine discussed earlier was developed in 1893 and remains a successful treatment for bladder cancer. The treatment uses a live bacteria injection to stimulate an immune response, which in turn reduces the tumor burden, although the exact mechanisms of the regression are not known. What is known is that the BCG vaccine attracts immune cells—CD4 and CD8 T cells, macrophages, and NK cells—to the location of the BCG injection.

Today, additional approaches include extraction of tumor cells from a patient in order to modify the cells to express the protein GM-CSF. This is intended to attract dendritic cells that would, "scurry back to regional lymph nodes and activate helper T-cells that can then launch a potent anti-tumor immune response," as per Weinberg [137].

There has been significant progress in cancer vaccines, though there is the hope of even greater progress. The FDA has approved vaccines targeting viruses that are known to cause cancer, such as liver cancer (caused by hepatitis B virus) and cervical cancer (caused by human papilloma virus). There is also Provenge (sipuleucel-T), which was approved in 2010 for metastatic prostate cancer. It is designed to stimulate an immune response to a specific protein (PAP) found on most prostate cancer cells.

Immune System Therapy—Passive Immunization

The passive immunization approach, as described by Weinberg [137], bypasses the immune system response and uses antibodies to attack an oncogenic protein directly. Herceptin (trastuzumab) is an example of such an antibody.

HER2 protein receptors are often upregulated in cancer. To target the hyperactive HER2 receptors, researchers produced antibodies in mice that could bind to and inhibit the human receptor. This approach opened a new avenue to treatment of breast cancer. A murine antibody, however, could not be used directly in humans because the human immune system will recognize the constant region of the mouse antibody and reject it. Therefore, the mouse antibody was genetically engineered by recombinant DNA techniques so that it would not be recognized as foreign by humans.

To do this, geneticists identified and cloned the gene that encodes the human constant region of antibodies. The human constant region was then fused with the gene encoding for the mouse variable region, which in turn recognizes the HER2 receptor. The result was the creation of the humanized or "chimerized" antibody, Herceptin, which could be produced easily in mouse cells and used to inhibit the oncogenic HER2 protein in patients while evading a response from their immune systems.

Although the mechanism by which Herceptin kills HER2-overexpressing breast cancer cells is controversial, there is no question that this occurs. Indeed, there are many monoclonal antibodies that have been developed and approved by the FDA. A partial list of antibody cancer drugs approved for use in humans is shown in table 15.1 [138].

Inhibiting Cancer-Causing Genes (Oncogenes)

Another approach to identifying the targets of therapy is to inhibit the expression of target genes using short double-stranded RNAs. Small interfering RNA (siRNA) technology can be targeted to halt the expression of genes known to drive cancer.

Table 15.1
Partial list of antibody cancer drugs approved for use in humans

Year Approved	Brand Name	Generic Name	Type of mAb	Indication
1997	Rituxan	Rituximab	Chimeric	NHL
1998	Herceptin	Trastuzumab	Humanized	Her-2/neu positive breast cancer
2000	Mylotarg	Gemtuzumab ozogamicin	Humanized	Acute myelogenous leukemia
2001	Campath-I	Alemtuzumab	Humanized	B-cell CLL
2002	Zevalin	Ibritumomab tiuxetan conjugated	Murine	NHL
2003	Bexxar	^{131}I-Tositumomab (radiolabeled)	Murine	NHL
2004	Avastin	Bevacizumab	Humanized	Colorectal cancer
2004	Erbitux	Cetuximab	Chimeric	Colorectal cancer
2006	Vectibix	Panitumumab	Humanized	Colorectal cancer
2009	Arzerra	Ofatumumab	Humanized	Lymphocytic leukemia

siRNAs inhibit the expression of target genes by base pairing with the target mRNA and causing its cleavage and subsequent degradation. The siRNA molecules, however, are large and extremely hydrophilic and are difficult to deliver to the intended cells. Moreover, once inside of the targeted cell, the siRNAs are vulnerable to degradation by nucleases [139].

One strategy that is currently used by many researchers is to package the siRNAs inside polymeric nanoparticles and decorate these with the polymer polyethylene glycol (PEG), which the body accepts as inert and helps to delay the clearance to kidneys or liver [139]. Although this strategy significantly improves the half-life and stability of the siRNA, it does not solve the problem of directed targeting to cancer cells. Thus, effective and efficient siRNA delivery remains an open engineering challenge.

Protein 53 (p53) restoration therapy has been a focus for 20 years [140]. p53 is the guardian of genomic stability, and a mutated p53 gene is found in approximately 50% [141] of all cancers. The attractiveness of restoring p53 function is obvious, but demonstrating efficacy and safety (this gene therapy has not been FDA approved) is problematic.

Another gene alternation therapy that has developed in recent years is intended to persuade the undifferentiated leukemic blast cells of acute promyelocytic leukemia (APL) to return to wild-type behavior and differentiate properly. Leukemias arise when the normal process of differentiation from a stem cell to a differentiated, highly specialized white blood cell is interrupted. In the case of APL, the process of making white cells called neutrophils stops at the promyelocyte stage [142]. These cells exhibit uncontrolled proliferation and crowd out the healthy cells in the marrow. Blast cells can be induced to become neutrophils, and thus cease to proliferate, by treatment with all-*trans*-retinoic acid (ATRA) and arsenic.

The discovery of ATRA and arsenic as a treatment for APL was not a result of the identification of a target protein and then development of an agent that acts upon the protein. Rather, it grew out of usage [143] to treat acute myelogenous leukemia.

The mechanism of action was only elucidated after its approved use in the clinic. The fusion gene PML-RARα produces a protein, which is believed to affect transcription controls and stimulate growth. Among the effects of the new gene is the reduction of PML (promyelocytic leukemia) proteins [144], which provide growth regulation and promote apoptosis. ATRA binds to RAR (retinoic acid receptor) receptors causing a degradation of the PML-RAR.

Engineering of Regulatory Pathway Inhibitors

With the developing understanding of pathways that underlie cancer growth and metastasis, the therapeutic approaches for cancer are increasing in great

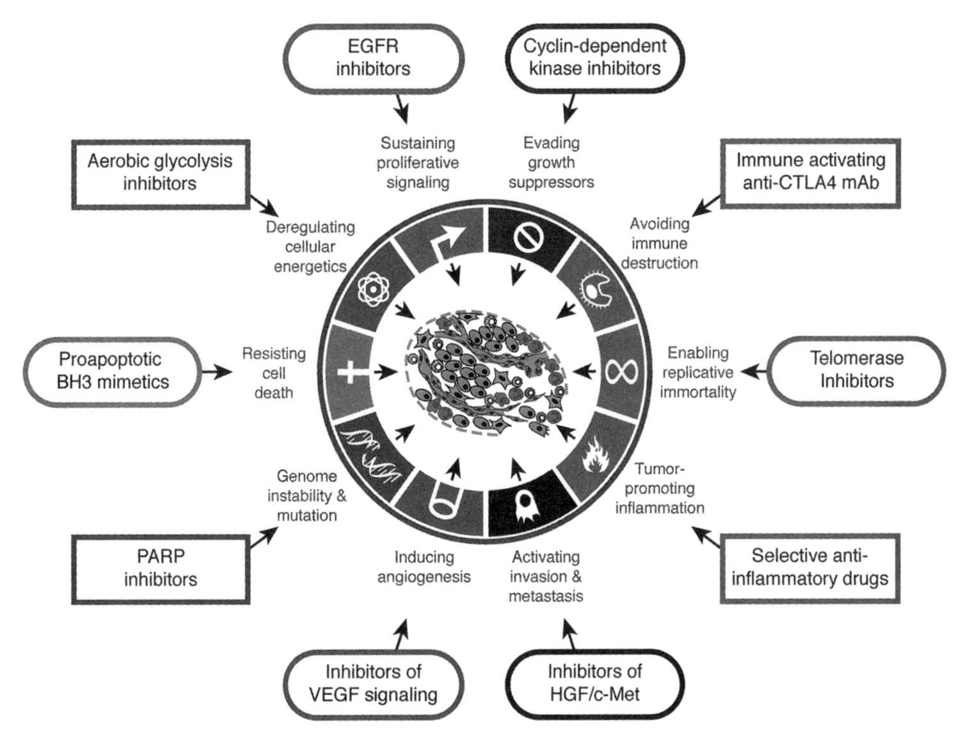

Figure 15.2
Therapeutic targeting of the hallmarks of cancer.
Source: From Douglas Hanahan and Robert Weinberg, "Hallmarks of Cancer: The Next Generation," *Cell* 2011; 144(5): 646–674. Reprinted with permission from Elsevier. Copyright © 2011 Elsevier.

numbers. Weinberg and Hanahan provide a nice overview of the various pathway-targeting drugs that have been developed or are likely to be developed for controlling cancer. Figure 15.2 shows the therapeutic targeting of the hallmarks of cancer.

Each of these targets for controlling cancer will require a deep and accurate understanding of the biology behind the regulatory pathways in cancer cells. Currently, the ability to control a dysregulated pathway in cancer requires extraordinarily time-consuming experimental programs to identify the activity of various components in the pathway. However, unlike the work on DNA sequencing, these pathways are multifocal processes, which means that more than a single measurement of a particular pathway has to be made. The problems in today's cancer research are fundamentally different from how science was done in prior centuries and will, no doubt, require new technologies.

Good targets have active regions that can be inhibited with small-molecule drugs. For example, among enzyme or receptor targets, a well-defined catalytic cleft or ligand-binding site has in the past led to functional small-molecule therapeutics, a state called "druggable." Experience has shown that kinases (enzymes) are involved in many of the oncogene pathways and are generally druggable. The downside is that the clefts are similar for many (tyrosine) kinases making it hard to create a specific drug that will not affect other pathways.

To understand better the new stage of cancer biology research and drug development, this book examines a case study that led to a new drug. The author will focus on the BRAF pathway and its use in treating melanoma.

Case Example: BRAF

In the late 1970s, the work of Varmus, Bishop, and others opened up the exploration of the regulatory processes controlling oncogenes and the propagation of cancer cells. By the turn of the decade, a number of key genes and their basic regulatory processes were becoming understood.

For example, in 1980 at the National Cancer Institute, J.R. Stephenson [145] and his team (Rapp, Goldsborough, Mark, Bonner, Griffin, and Reynolds) were investigating the characteristics of cancer induced by a mouse type C virus. In particular, they were researching its apparent ability to express an unknown cancer-causing gene. The team induced cancer in mice with the virus and isolated the gene that caused oncogenic transformation. Using Southern blotting, they found that it bore no resemblance to previously identified oncogenes and named it V-RAF (RAF stands for rapidly accelerated fibrosarcoma [first identified by Rous] and V stands for virus-induced.

At the same time, Tony Hunter, William Raschke, and Bartholomew Sefton at the Salk Institute were working with the cancer-causing gene also from the Rous sarcoma virus [146] when they found that its ability to phosphorylate other proteins at tyrosine residues was required for its capacity to induce transformation. The protocol used [147] was

- virus infection of cells labeled with radioisotopes;
- immunoprecipitation of cell lysate;
- gel electrophoresis and subsequent peptide mapping with radioactive polypeptide detection by fluorography of the phosphorylated protein.

Working on the Abelson murine leukemia virus, they discovered the pathway of the virus in which the virus encoded a protein kinase that phosphorylates tyrosine in the transformed cells. This, they wrote, "Suggest[s] by analogy that the modification of cellular polypeptides through the phosphorylation of tyrosine may be involved in cellular transformation by Abelson virus." Thus, the role of phosphorylation of tyrosine residues in proteins by a protein kinase came into focus [148].

About 13 years later, in 1993, work was completed by Deborah K. Morrison [149] and her team on characterizing the activity generated by the mutant gene. By treating cells with several growth factors, they observed "The close correlation between Raf-1 phosphorylation and kinase activation" of the RAF kinase.

They used immunoprecipitation, followed by peptide mapping, then reverse-phase high-performance liquid chromatography, and finally analysis in a (Beckman) protein sequencer. Further work identified that the two variants of RAF, A-Raf and B-Raf, both share the same three conserved domains [149].

The technologies used to discover the mechanism of kinase function were

- Immunoprecipitation
- Radiolabeled gel electrophoresis (1937)
- Analysis by peptide mapping using gel electrophoresis (1937)
- Analysis by protein sequencer (Beckman 890D, 1969)

The evolving picture of the RAF genes and their role in proliferation unfolded over many years leading to an understanding of their role in complex signaling pathways. One such pathway is shown in figure 15.3 for illustrative purposes. In this prototypical pathway, a surface binding event (EGF at the

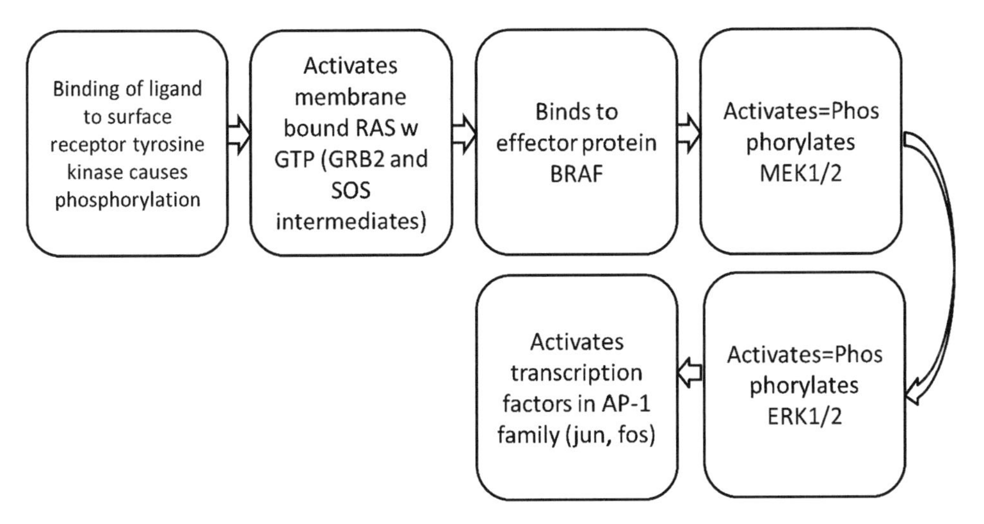

Figure 15.3
MAPK signaling pathway.

EGFR) activates a receptor, which in turn activates the protein RAS (rat sarcoma) and then BRAF (the marine sarcoma viral oncogene homolog B1 of RAF gene). BRAF then activates MEK (mitogen-activated kinase) and ERK (extracellular signal-regulated kinase), which activate transcription factors. The transcription factors trigger the expression of genes that promote cellular proliferation. A mutated RAS and/or BRAF gene continues this cycle without stop; this creates the tumor.

In 2002, Michael R. Stratton, P. Andrew Futreal, and colleagues published their results demonstrating that BRAF mutations were at the root of melanoma cancers [150]. Of the 43 probable BRAF somatic mutations, one mutation (BRAF V600E) was found in most cancers.

The next step in the unraveling of the BRAF process of cancer came in 2004 with the work of Richard Marais [151] when he isolated the active proteins of the mutant BRAF V600E chain and performed X-ray diffraction studies to characterize the active region of the protein. This yielded a complex three-dimensional characterization of the BRAF V600E kinase, which would become the target for drug development.

Work had been going in several laboratories to identify a small molecule that would inhibit the mutated BRAF protein. Among them were several drug company teams that converged on the BRAF V600E mutated kinase active region. The drug SB-590885 by GlaxoSmithKline [152] inhibited the active region BRAF V600E by binding specifically in four locations and produced a 17,000-fold selectivity against other kinases.

Plexxikon developed PLX403 in 2008 [153], another inhibitor to the mutated BRAF kinase. The pipeline includes GSK2118432, PLX4720, RAF265, XL281, AZD6244, and GSK1120212. These inhibitors had terrific results shrinking tumors and increasing survivability [154]. The survivability of patients improved an average of 7 months (range of response from 2 to more than 18 months).

It is, however, not a cure from melanoma, meaning that the person dies of the same disease, which finds a way to progress by alternative mechanisms. Resistance to the BRAF inhibitor develops in nearly all cases, and the cancer returns. Several alternative pathways to the mutant BRAF have also been established [155]. However, as the researchers conclude, "Melanomas escape B-RAF(V600E) targeting not through secondary B-RAF(V600E) mutations but via receptor tyrosine kinase (RTK)-mediated activation of alternative survival pathway(s) or activated RAS-mediated reactivation of the MAPK pathway, suggesting additional therapeutic strategies."

BRAF and its importance to cancer was first identified around 1987 [156], and in 2008 the first drug was developed. Some 20 years of research and development were dedicated to the inhibitor. Based on recent clinical results, the 20-year effort addresses about one-third of the pathways of the mutated BRAF, and so much work still lays ahead to close down these alternate pathways. Clearly, discovery of the tumor-causing pathway and the alternative pathways that cause apparent resistance to the drugs must be accelerated. We cannot wait 20 years for each individual cancer pathway to be characterized. We believe that new technologies for the rapid characterization of cancer pathways (also discussed in chapter 14) is an huge unmet need.

16 Summing Up

As we traveled through the numerous examples of engineering-enabled science, we have seen some common phenomena. Tracing back to the 1800s, we see in the world of physics—from subatomic to galactic—that scientists conceived, designed, and, in some cases, built their own technology. They did this so they could pursue experiments to explore the heart of the atom or the universe and understand its composition and structure.

Famous physicists such as Ernest Rutherford, Robert Andrews Millikan, Albert Abraham Michelson, and others not only understood the physics of matter but also had the conceptual talent to engineer devices and instruments to enable their breakthrough discoveries. They were self-sufficient scientists; engineers capable of embarking on a scientific investigation and building the technology along the way.

Consider table 2.1 in chapter 2. For seven discoveries in physics, there were seven de novo technologies built. Over the past decade or so, many discoveries in physics have come from shared technologies such as the Hubble Space Telescope and accelerators. And even still, things such as the discovery of giant magneto resistance (where new structures were developed) and the Bose–Einstein condensate (where a cryogenic apparatus was built to achieve 170 nanokelvin), just to name two of the many recent examples, were enabled by technology built exclusively for the purposes of the experiment.

Of 40 Nobel Prizes in Physics, most were enabled by purpose-built apparatuses that were necessary for the science. A number of these experiments generated new data—structure of the nucleus, confirmation of quantum theory and wave theory of matter, to name a few—that changed science paradigms. Thomas Kuhn describes this as the nature of scientific advancement.

As we mentioned earlier in the book, Leroy Hood has said, “Always practice biology at the leading edge of biology and [invent] develop a new technology for pushing back the frontiers of biological knowledge.” Again in physics, we saw in chapter 2 that the tight integration of engineering hardware to support scientific investigation continued and expanded through the twentieth century.

During the early to mid-1900s, the scope of technology development in physics increased, and larger teams were necessary to undertake and complete the development of the hardware. The evolution of larger integrated teams of engineers and scientists exploring high-energy physics gave way to the development of the national laboratories. This also led to the creation of some of the most spectacular engineered technology to support science. The accelerator technologies at the Lawrence Radiation Laboratory and at Berkeley and Livermore as well as Brookhaven and the Stanford linear accelerator are all classic examples of integrated efforts of engineers and scientists to develop technology to enable the exploration of the physical sciences.

In the case of the Hubble Space Telescope, concurrent science and engineering was achieved in no small measure by the work of one man, Dr. Rodger Doxsey (see chapter 2), who successfully straddled both worlds.

Recently, there has been growing attention given to the convergence of engineering and science [157]. In reality, it has been practiced in a number of the sciences for over two centuries and in biology during two periods of time: from 1850 to 1900 and from 1950 to 1970. In the early days, before electronic communication and modern scientific infrastructure, it was largely accomplished by very small teams or single individuals capable of doing the engineering and the on their own. That of course represents the ultimate convergence of the disciplines and is something that our current educational system has not yet embraced broadly.

40 Years of Evolutionary versus 10 or Less Years of Concurrent Technology Development

What is the value of concurrent technology development and research?

Consider what we have learned about discovery in the science of biology from Mendel's work through the twentieth century and the unraveling of the mysteries of DNA. Until around 1950 and the molecular biology revolution, the major milestones in biology (as measured by Nobel Prizes) were enabled by technologies that had existed but were not made available to the biologist for an average of 40 years (figure 16.1).

By contrast, there was the leading-edge biology conducted concurrent with technology development by well-known leaders Sanger, Perutz, Moore and Stein, and others. Their concurrent technology development during that period was not as broadly known but was certainly important as exemplified by Phillip A. Sharp's comments in chapter 11.

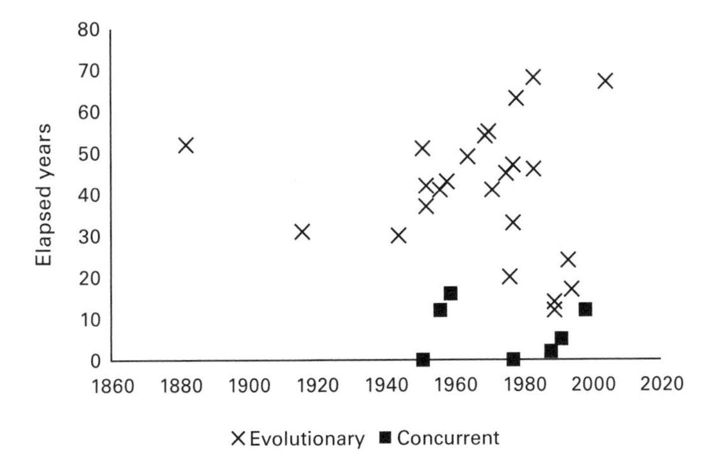

Figure 16.1
Years elapsed from the time of development of an enabling technology (proof of concept) to when it was used in breakthrough biological discovery. Starting around 1950, concurrent engineering supporting biological discovery helped to accelerate discovery.

We observed that in most cases, nearly four decades elapsed for a new physical mechanism to be engineered into a device that the broader community could use for research. Once available the technology appears to have enabled Nobel Prize–winning discovery in about a decade.

Kuhn sheds some light on how this can happen when he observed that scientific knowledge comes with leaps and new paradigms and is "non-cumulative developmental episodes, in which an older paradigm is replaced in whole or in part by an incompatible new one" [78]. Perhaps the discovery is not paced by the cumulative scientific knowledge as much as by the availability of new data from the new technology. It is as though the notion to explore a particular line of biological investigation is pent up in the research community and is aggressively pursued when a new enabling technology is made available, thereby leading to the breakthrough.

The author wonders if, in some cases, the scientist had some original ideas for investigation but disregarded them because the technology was not there to support exploration. Research was not taken up again until when the enabling technology became available. This is consistent with some of Dr. Ron Davis's observations in chapter 11.

Advancements in biology contain many wonderful examples of new data (X-ray diffraction of DNA helix) leading to the "ah-ha" moment (Crick and Watson), which created a leap in understanding and a paradigm shift in the science. When we unravel the story of the DNA helix, it traces back to the discovery of physical phenomenon (X-ray diffraction), the long development time to engineer the mechanism into a useful tool, and then the biological discovery—two ah-ha moments that bound the timelines of our study.

We did not find the same timeline profile for technologies that addressed clinical needs. The end points of these stories are the release of the technology to the commercial world and entry into the clinic. These tend to be more evolutionary in nature and driven by a compelling unmet need. Gelijns [103] observes in her work that the "effective coupling of engineering to clinical medicine, speeds the introduction of new diagnostic and therapeutic interventions."

This book was written largely to explore concurrent engineering and basic biology discovery and is in no way a complete story. The author included the chapter on enabling technologies for the clinic as a means of comparison.

Epochs of Convergence in Biology, 1850–1900 and 1950–1970

There appear to be two time periods in biology when engineering and biology were done concurrently. The first is the period 1850–1900 when biochemistry developed, and the work of Pasteur, Fourcroy, Mulder, Miescher, and others formed the fundamental understanding of key biological processes. At the same time, light microscopy and spectral analysis provided a huge leap in seeing the composition of cells, proteins, and DNA. However, the function of these elements was not understood until the next major epoch of discovery started in 1950.

The next big leap in combining technology and biology occurred in the 1950s when molecular biology was driven at the Rockefeller Institute, the Laboratory for Molecular Biology, and Caltech by such notables as Max Perutz, Moore and Stein, Sanger, Tiselius, and others. A key factor in their success was their respective abilities to develop leading-edge technology to support their biological research.

Note in figure 16.2 the concentration of discoveries in the period 1950–1960 and the long gestation time leading up to the enabling technology development availability (left side of the shaded block).

Today, the time delays of the past can be decreased by the dramatic increase in the sharing of research that has come with our information age. It is now possible for a researcher to browse the Internet freely to discover the seeds of early technologies that might be applicable to a research program and create collaborations for concurrent engineering and biology.

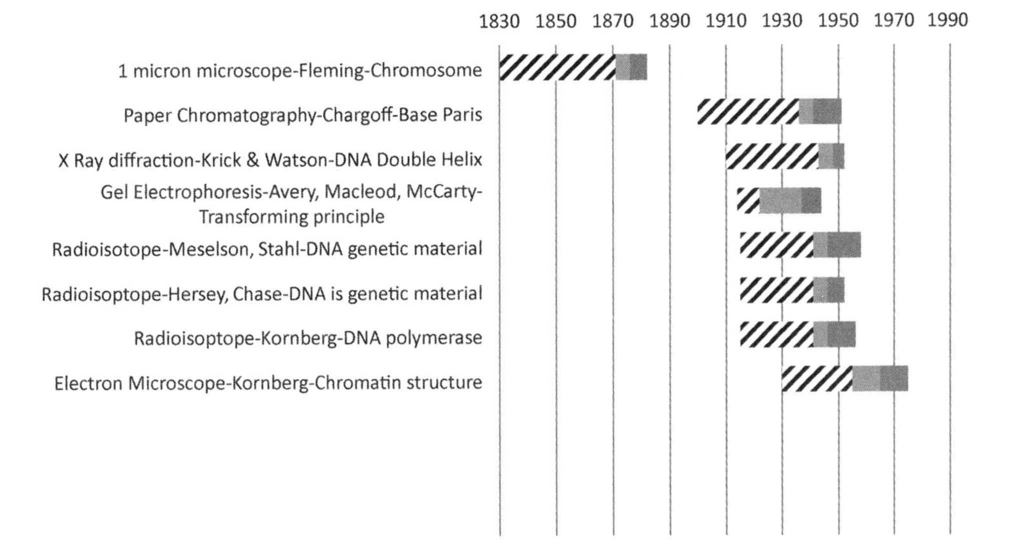

Figure 16.2
Enabling technology timeline: earliest box (leftmost) is elapsed time from technology development (proof of concept) to the commercial development, then duration of commercialization (middle box), and finally (in gray) is the time to published biological discovery. Note the concentration of discoveries in 1950 to 1960 and the long gestation time leading up to the enabling technology development and availability (the two blocks to the left).

Six Key Enabling Technologies Support Many Biological Discoveries

The 37 molecular biology and genetics discoveries examined in this book were enabled by six fundamental technologies. At the heart of the genetics and molecular biology discoveries of the twentieth century were microscopy (chapter 4), gel electrophoresis and chromatography (chapter 5), X-ray diffraction (chapter 6), radioactive tracers (chapter 7), electron microscope (chapter 8), and sequencing (chapter 9).

These technologies and their respective stories compare to the 30 purpose-built technologies that supported 40 discoveries in physics along with the large shared technologies of the Hubble Space Telescope and accelerators.

Assuming this sampling is representative of the whole field, it is interesting to note that in the past few decades, nearly 70% of U.S. academic researchers are classified as working in biology while 5% are in physics. The implication is that there is a lot of biological research using undifferentiated available technology to generate data with probably limited opportunity to discover data that would cause a paradigm shift. As Jeffrey Drazen characterized the situation, "Our thinking is short sighted and linear. For example, as an immunologist graduate student you spend years learning all of the facts. You understand the kinds of experiments from which the facts were derived. As a faculty member it is your turn to do experiments, so you do experiments just like the ones from which the current facts arrived in order to derive a new fact. You never think beyond that horizon."

Gel electrophoresis and chromatography both rely on differential transport through a medium for separating and studying proteins, DNA, and other large molecules. Sanger sequencing depends upon gel electrophoresis so it falls into the same category. These technologies can be thought of as media separation processes. Of the 37 breakthroughs in biology we looked at, 15 of these were enabled by gel electrophoresis, blots, or chromatography; three by X-ray diffraction; eight by radioisotope labeling; two by light microscope; three by electron microscope; three by PCR; two by sequencing; and two by other technologies (spectrometer, ultracentrifuge).

Nobel Prize Winners Tend to Be Early Technology Adopters

We note that even when the enabling technology is developed commercially and not concurrently with the biological research, the Nobel Prize winners in biology tend to be early adopters of the enabling technology. They average about 12 years between the time the new technology is available and when they published their groundbreaking results. Meanwhile, the limited sampling of other researchers included in our analysis averaged 30 years.

Ingredients for the Leaps in Biology

Four factors behind the biology breakthroughs were extracted from the research of the cases in this book. The first and most obvious is the availability of enabling technology to open the new line of thinking and research. Engineering the enabling technology concurrently with the biological research accelerates the time to discovery or translation to the clinic. Three institutional factors appear to be necessary to create effective concurrent engineering and biology: a peer group that is a mix of disciplines, a visionary who creates the setting and points to the enabling technology, and funding to support the integrated (larger) teams.

Enabling Technology

As shown (the author hopes) throughout the book, concurrent engineering and biology creates new technology, which enables the next generation of new data and insights.

Peers of Mixed Disciplines

Shortly after Sydney Brenner learned that he and two former labmates had won the 2002 Nobel Prize, he received this e-mail from a Chinese researcher: "I wish also to win a Nobel Prize. Please tell me how to do it." The answer, Brenner announced at the award ceremony, is simple: "First you must choose the right place . . . with generous sponsors to support you." In addition, he urged, "Choose excellent colleagues."

For Brenner and a dozen other Nobel laureates, the right place was Cambridge, England, and the right people were their peers at one of the world's first laboratories devoted to molecular biology [158].

James D. Watson has called the Laboratory of Molecular Biology (LMB) "The most productive center for biology in the history of science." The laboratory's recipe for success, he added, is that "The LMB was able to concentrate in one place very exceptional scientists. . . . Today, some of that talent would probably not make the first cut for a university position, given the apparent discrepancy between the scientist's experience and the job description. Sulston, for instance, was a chemist working on the origins of life when he came to study the worm. Max [Perutz] had this uncanny ability to see something special" [158].

As discussed earlier, the peer support among the great researchers at the LMB has been chronicled. Throughout the biography of Perutz [40], Georgina Ferry described the formal and informal interaction across the disciplines of the members. Critical thinking and questions among peers led to out-of-the-box thinking and discoveries.

Similarly, there was a memorable moment in the interview [92] of J. Leighton Read. When reflecting on the moment of invention of using photolithography to create microarrays, he said that the cross-discipline group that Alex Zaffaroni pulled together was critical for preventing the seed of a novel approach from "Falling on fallow ground."

Another interesting finding was captured in the interview with Phil Sharp. His physical chemistry background gave him knowledge, techniques, and motivation to explore the leading edge of genetics, amplifying the possibility of important discoveries. In contrast, Ron Davis—in Norman Davidson's laboratory at Caltech at the same time as Sharp—was a geneticist first and created the environment to invent new technologies to support his biology work. One can come at the combination of concurrent technology development and biological discovery from one or the other discipline.

As we explored in chapter 14, the development of medical products typically has a much closer coupling between the engineering of the technology and its use in medical practice. Still as Gelijns [103] points out, the close coupling of engineers to clinical medicine speeds the introduction of new diagnostics and therapeutics. The products were developed to meet a medical need and therefore upon completion were used as rapidly as the developer could deliver them.

As noted earlier, biology and technology discovery tend to occur in separate siloes, though this paradigm may be shifting due to information sharing enabled by the Internet. However, to overcome this challenge, the culture, reward system, and peers must contribute positively in supporting the scientist exploring an anomaly (Kuhn) so that it may be the basis of a new discovery. As Jeffrey Drazen points out, "Remember it was back in 1900 when Lord Kelvin said that there was nothing new to be discovered in physics. It wasn't until Einstein came out of the woodwork with some ideas we hadn't had before that Kelvin was proved wrong."

Visionary

In the case of Caltech and the LBM, leadership brought more than management but also a vision and made that vision a reality.

At Caltech, Thomas Hunt Morgan set out to create a genetics powerhouse. Similarly, Max Perutz (and now Leroy Hood) saw the importance of technology development to push the boundaries of biology.

Both of these stories—the institutions and their leadership—underscore Jeffrey Drazen's comment quoted in chapter 12,[1] "Your (the author's) premise is that the really big progress in biology comes when there is somebody who is visionary enough to say 'If I can measure something . . . ,' and then if he or she is able to work with somebody who helps [provides the expertise] to help measure it. Then they are able to get into an area where no one has ever been before, they discover new ideas. Pretty soon, as people realize how important these measurements are, the instrument companies get into the business of making the technology available commercially."

> So you need to have somebody, for example such as Perutz. Somebody who is running the lab who is smart enough to figure out where the problems are and what type of technology is needed to solve them. At the Laboratory of Molecular Biology it was X-ray crystallography. So where they had all those people making the needed instruments and learning to read the diffraction patterns and so on . . . . Suppose Perutz hadn't found the right people to do that. What if he hadn't had that insight, little would have happened.

So the brilliance is to figure out what the general type of technology needs to be invented to solve the unposed problems. So when the biologist hangs out with the physical scientist and the engineer, they can understand each other's problems enough to help invent the technology solutions.

In practice, we don't really know where the ideas are going to come, from but it is a matter of creating an environment where people who have ideas can explore them in a cross-disciplinary fashion.

As mentioned numerous times in this book, Leroy Hood recalls his mentor Bill Dreyer's (at Caltech) guiding principle that it is important to "Always practice biology at the leading edge of biology and [invent] develop a new technology for pushing back the frontiers of biological knowledge" [70].

Ron Davis (director of the Stanford Genome Center) gives us some insight into how he develops approaches and waits until they are useful: "There is a factor of time. It isn't a matter that you make this list and sit down one day you try to match them up, right? This is an ongoing process and you have these solutions that you don't have the right problems for. It might be ten years later that you run into this problem and you say, 'Ah! We figured out the solution to these ten years ago.' You keep them. You make sure you integrate that into time; and the larger your list of solutions and problems, the more likely you get hits. And the longer the time frame increases your chances."

This is very common for us to come up to a problem we are going to take and we have the solution. It just hits you perfectly. . . . People had the idea and tried to work on it, but the technology wasn't there. As soon as the technology comes in, the next person who thinks of it is the one who is going to get the Nobel Prize.

Thomas Kuhn believes that scientific advancement occurs via a result that was an anomaly [1] of some type where the scientist(s) had the insight to recognize the inconsistency with the current paradigms. While this may come from repeated measurements with the same six major biology tools, history has shown that the big leaps come from new data generated with new tools.

Funding

Visionaries need funding. Most of the breakthrough biology we looked at arose from special and unusually continuous funding. Perutz and Sanger supported by the Medical Research Council (Sir Harold Himsworth in particular) and Rockefeller (Dr. Gerard R. Pomerat); More and Stein by Rockefeller; Leroy Hood and special funding from Science and Technology Center grants (National Science Foundation); and Eric Lander getting private funding for the Broad Institute.

Without the benefactor, the researcher must put together a patchwork of funding sources, which is a methodology that is unlikely to ever support a concurrent development of any scale. Roger Kornberg said it well, "I had to develop creative funding strategies as no single grant would cover the entire scope of stretching from detail structural chemistry to biological processes."

One of the strategies he describes is the need to define interim objectives, which can attract funding: "One can then do related work and draw upon the funds to explore the basic science. It is understood you can use the money for other purposes, but the work must be successful." This approach of course limits any work in new enabling technologies to a sliver of the funding pie.

There is a dichotomy. The vast majority of NIH funding is distributed in relatively small R01 grants (small science). Individual investigator, R01-style grants for smaller projects are a time-honored and respected tradition. The thinking is that the 30,000 research project grants—mostly R01—made annually build a body of knowledge that cumulatively move our understanding along.

However, as Kuhn points out, the major steps in science are "non-cumulative" and depend upon new data and are largely not dependent on the current understandings. As this book points out, new data are often created from new technology, which cannot be developed concurrently in small individual R01 style projects. This is because of scale limitation and also the difficulty in funding these through the peer review process.

Therefore, are we doing the right thing when it comes to funding methodology?

Based on the historical perspective of what it takes to have leading-edge biology that achieves the advancements we all wish for, the author believes there should be a tangible change to the NIH funding model. Along the lines of the Cook-Deegan article (chapter 13), NIH could benefit from the integrative technology development program expertise at DARPA. A number of biology-oriented program managers from DARPA could:

- Interview NIH directors and managers to develop a high-level view of the unmet needs, provocative questions [159], and major challenges.
- Conduct life sciences workshops and field interviews on the needs and challenges to encompass what is known and articulate the requirements for what is needed. Determine the types of concurrent engineering and biology projects that would be important and achievable.
- Assist NIH in the procurement process. Special emphasis panels or a new study section could be used.

Next Visionary, Next Technology

Who and what will be the next visionary and technology? We traveled through the 1950s and X-ray diffraction, 1960s and electron microscopy, and the 1990s and genome sequencing. Will it be cell signaling, pathways, systems biology?

We do not know if there are any master controls on cell signaling or gene expression. Some people believe that metabolic pathways could be master controls. However, to determine if this is the case, we need new technology to help us "see" the dynamics of controls in the cell.

The Hubble Space Telescope provided a technological means of seeing things that could not under any circumstance be seen by any technology on Earth. We need similar technology innovations to "see" mechanisms of cell signaling in real time and in vivo in order to observe the intact biology. We need a Space Telescope of Biology. Similarly, and for example, we need detection and localization of cancer tumors prior to vascularization (less than 1 mm) and ex vivo organ models that recapitulate human systems, including immune response.

Somewhere there may be a nascent technology in some physics or electronics or computer science laboratory that could reveal new data on cell networks (signaling and pathways). Somewhere else there may be our visionary formulating the next LMB to generate the new data and open the paradigm shift that will make tractable our understanding of cellular networks, which seem so daunting now.

Will we have to wait 40 years for this to come together?

Notes

Chapter 12

1. See Lander's interview on the development of genome sequencing in chapter 11.

Chapter 13

1. In 1944, Rabi was awarded the Nobel Prize in Physics "for his resonance method for recording the magnetic properties of atomic nuclei."

Chapter 14

1. The name is derived from the 1914 observation by John S. Dexter of a notch in the fruit fly wing, which was studied over the next five decades to identify the molecular controls.
2. See timelines in chapter 10.

Chapter 16

1. Drazen was provided with a copy of the manuscript of this book prior to our interview.

References

1. Kuhn, T. S. 1962. Anomaly and the emergence of scientific discoveries. Chapter VI in *The Structure of Scientific Revolutions*. Chicago: University of Chicago Press.

2. Marsden, H. G. E. 1913. The laws of deflexion of alpha particles through large angles. *Philosophical Magazine* 25 (148):1913.

3. Karlsson, E. B. The Nobel Prizes in Physics 1901–2000. Available from: http://www.nobelprize.org/nobel_prizes/physics/articles/karlsson/.

4. Richmond, M. The Franck-Hertz experiment supports Bohr's model. Available from: http://spiff.rit.edu/classes/phys314/lectures/fh/fh.html.

5. Lawrence Berkeley National Laboratory. Lawrence with the 37 1/2 inch cyclotron. 1935. Lawrence Berkeley National Laboratory.

6. Lawrence, D. Noble Prize lecture. Available from: http://nobelprize.org/nobel_prizes/physics/laureates/1939/lawrence-lecture.pdf.

7. Lab staff standing on the 184 inch cyclotron. Available from: http://www.physics.rutgers.edu/cyclotron/images/cyc_hist_album/images/cyc_hist_26.jpg.

8. Berkeley Lab HILAC, August 1965. Available from: http://www.lbl.gov/image-gallery/image-library.html.

9. Overbye, D. Rodger Doxsey, astronomer who worked on the Hubble, dies at 62. Available from: http://www.nytimes.com/2009/10/19/science/space/19doxsey.html.

10. The Free Dictionary. Available from: http://www.thefreedictionary.com/engineering.

11. *Merriam-Webster's Dictionary*. Available from: http://www.merriam-webster.com/dictionary/genetic%20engineering.

12. MIT Biological Engineering Department. Available from: http://web.mit.edu/be/index.shtml.

13. Chandler, D. L. Going head to head. *MIT News* February 9, 2011.

14. Adami, H. O. 2009. Epidemiology and the elusive Nobel Prize. *Epidemiology (Cambridge, Mass.)* 20 (5):635–637.

15. Henig, R. M. 2001. *The Monk in the Garden: The Lost and Found Genius of Gregor Mendel, the Father of Genetics*. Houghton Mifflin.

16. Perrett, D. 2007. From 'protein' to the beginnings of clinical proteomics. *Proteomics. Clinical Applications* 1 (8):720–738.

17. Mulder, G. J. 1796. *The Chemistry of Vegetable and Animal Physiology*. William Blackwood and Sons.

18. Cahan, D. 1996. "The Zeiss Werke and the Ultramicroscope: The Creation of a Scientific Instrument in Context." In *Archimedes: Scientific Credibility and Technical Standards in 19th and Early 20th Century Germany and Britain*, Jed Z. Buchwald, ed. Vol. 1. pp. 67–115. Dordrecht: Kluwer Academic.

19. Flemming, W. 1878. Zur Kenntniss der Zelle und ihrer Theilungs-Erscheinungen. *Schriften des Naturwissenschaftlichen Vereins für Schleswig-Holstein* 3:23–27.

20. Michael, W. and P. Davidson. Ernst Abbe microscopy. Available from: http://labmed.ascpjournals.org/content/40/8/502.full.pdf+html.

21. Kober, P. A. 1915. Spectrographic studies of amino acids and polypeptides. *Journal of Biological Chemistry* XXII (3):433–441.

22. Smith, F. C. 1928. The ultra-violet absorption spectra of uric acid and of the ultra-filtrate of serum. *Biochemistry Journal* 22 (6):1499–1503.

23. Lewis, S. J. 1917. *A New Sector Spectrophotometer*. RSC Publishing.

24. Foundation, T. N. The Nobel Prize in Physiology or Medicine 1933, Thomas H. Morgan. Available from: http://www.nobelprize.org/nobel_prizes/medicine/laureates/1933/morgan-bio.html.

25. Avery, O. T., C. M. Macleod, and M. McCarty. 1944. Studies on the chemical nature of the substance inducing transformation of pneumococcal types: Induction of transformation by a desoxyribonucleic acid fraction isolated from pneumococcus type III. *Journal of Experimental Medicine* 79 (2):137–158.

26. Perrett, D. 2010. 200 years of electrophoresis. *Chromatography Today*. Available from: http://www.chromatographytoday.com.

27. Brannigan, L. H. Theodor Svedberg. Available from: http://www.chemistryexplained.com/St-Te/Svedberg-Theodor.html.

28. Cohen, S. 1999. *Interview of William J Dreyer*. Archives California Institute of Technology.

29. McCoy, R., C. E. Meyer, and W. C. Rose. 1935. *Feeding Experiments with Mixtures of Highly Purified Amino Acids VII. Isolation and Identification of a New Essential Amino Acid*. American Society of Biological Chemists.

30. Sumner, J. B., and S. F. Howell. 1936. Identification of hemagglutinin of jack bean with concanavalin A. *Journal of Bacteriology* 32 (2):227–237.

31. Sumner, J. 1946. The chemical nature of enzymes. Nobel lecture. Available from: http://www.nobelprize.org.

32. Issaq, H. J. 2000. A decade of capillary electrophoresis. *Electrophoresis* 21 (10):1921–1939.

33. Magasanik, B., E. Vischer, R. Doniger, D. Elson, and E. Chargaff. 1950. The separation and estimation of ribonucleotides in minute quantities. *Journal of Biological Chemistry* 186 (1):37–50.

34. Sanger, F. 1988. Sequences, sequences, and sequences. *Annual Review of Biochemistry* 57:1–28.

35. Martin, A. J. P. 1952. The development of partition chromatography. Nobel lecture. Available from: http://www.nobelprize.org.

36. *Complete Dictionary of Scientific Biography*. 2008. Archer John Porter Martin. New York: Scribner.

37. Nobel Lectures, Physiology or Medicine 1942–1962. Biography: James Dewey Watson. Available from: http://www.nobelprize.org/nobel_prizes/medicine/laureates/1962/watson-bio.html.

38. Watson, J. 1968. *The Double Helix: A Personal Account of the Discovery of the Structure of DNA*. Touchstone.

39. The Nobel Prize in Chemistry 1962: Biography, Max F. Perutz. Nobel Lectures. Chemistry: 1942–1962. Available from http://www.nobelprize.org/nobel_prizes/chemistry/laureates/1962/perutz-bio.html.

40. Ferry, G. 2007. *Max Perutz and the Secret of Life*. Cold Spring Harbor Laboratory Press.

41. Kauzmann, W. 1956. Structural factors in protein denaturation. *Journal of Cellular Physiology. Supplement* 47 (Suppl 1):113–131.

42. Kauzmann, W. 1959. Some factors in the interpretation of protein denaturation. *Advances in Protein Chemistry* 14:1–63.

43. Muirhead, H., and M. F. Perutz. 1963. Structure of haemoglobin. A three-dimensional Fourier synthesis of reduced human haemoglobin at 5–5 a resolution. *Nature* 199:633–638.

44. Creager, A. N. H. 2009. Phosphorus-32 in the phage group: Radioisotopes as historical tracers of molecular biology. *Studies in History and Philosophy of Biological and Biomedical Sciences* 40:29–42.

45. Hanawalt, P. C. 2004. Density matters: The semiconservative replication of DNA. *Proceedings of the National Academy of Sciences of the United States of America* 101 (52):17889–17894.

46. Stahl, M. M. a. F. 1958. The replication of DNA in *Escherichia coli*. *Proceedings of the National Academy of Sciences of the United States of America* 44 (7):671–682.

47. Kornberg, A. Noble Prize lecture. Available from: http://www.nobelprize.org/nobel_prizes/medicine/laureates/1959/kornberg-lecture.html.

48. Hurwitz, J., J. J. Furth, M. Anders, P. J. Ortiz, and J. T. August. 1961. The enzymatic incorporation of ribonucleotides into RNA and the role of DNA. *Cold Spring Harbor Symposia on Quantitative Biology* 26:91–100.

49. Kornberg, R. D. Roger D. Kornberg—autobiography. Available from: http://www.nobelprize.org/nobel_prizes/chemistry/laureates/2006/kornberg.html.

50. Interview of Roger Kornberg. Available from: http://www.nobelprize.org/mediaplayer/index.php?id=78.

51. Wikipedia. Transmission electron microscopy. Available at: http://en.wikipedia.org/wiki/Transmission_electron_microscopy.

52. Riddle, L. Biographies of women mathematicians: Dorothy Maud Wrinch. Available from: http://www.agnesscott.edu/lriddle/women/wrinch.htm.

53. Pauling, L., and C. Niemann. 1939. The structure of proteins. *Journal of the American Chemical Society* 61:1860–1867.

54. Bud, R., and D. J. Warner. 1998. *Instruments of Science: An Historical Encyclopedia*. Garland Encyclopedias in the History of Science. Science Museum, London, and National Museum of American History, Smithsonian Institution, in association with Garland Publishing.

55. Edman, P., and G. Begg. 1967. A protein sequenator. *European Journal of Biochemistry* 1 (1):80–91.

56. Eskow, D. 1983. Here come the assembly-line genes. *Popular Mechanics*, March.

57. Kresge, N., R. D. Simoni, and R. L. Hill. 2005. The elucidation of the structure of ribonuclease by Stanford Moore and William H. Stein. *Journal of Biological Chemistry* 280 (50):e47–e48.

58. Kresge, N., R. D. Simoni, and R. L. Hill. 2005. The fruits of collaboration: Chromatography, amino acid analyzers, and the chemical structure of ribonuclease by William Stein and Stanford Moore. *Journal of Biological Chemistry*. 280:e6.

59. Hirs, C. H., W. H. Stein, and S. Moore. 1956. Peptides obtained by chymotryptic hydrolysis of performic acid-oxidized ribonuclease; a partial structural formula for the oxidized protein. *Journal of Biological Chemistry* 221 (1):151–169.

60. Stein, W. Autobiography. Available from: http://www.nobelprize.org/nobel_prizes/chemistry/laureates/1972/stein.html.

61. Rushizky, G. W., and C. A. Knight. 1960. A mapping procedure for nucleotides and oligonucleotides. *Biochemical and Biophysical Research Communications* 2:66–70.

62. Larner, J., and F. Sanger. 1965. The amino acid sequence of the phosphorylation site of muscle uridine diphosphoglucose alpha-1,4-glucan alpha-4-glucosyl transferase. *Journal of Molecular Biology* 11:491–500.

63. Min Jou, W., G. Haegeman, M. Ysebaert, and W. Fiers. 1972. Nucleotide sequence of the gene coding for the bacteriophage MS2 coat protein. *Nature* 237 (5350):82–88.

64. Gilbert, W., and A. Maxam. 1973. The nucleotide sequence of the lac operator. *Proceedings of the National Academy of Sciences of the United States of America* 70 (12):3581–3584.

65. Carr, D. S. M. Steps toward DNA sequencing: Maxim and Gilbert method. Available from: http://www.mun.ca/biology/scarr/4241_StepstowardsDNASequencing.html.

66. Smith, L. M., J. Z. Sanders, R. J. Kaiser, P. Hughes, C. Dodd, C. R. Connell, et al. 1986. Fluorescence detection in automated DNA sequence analysis. *Nature* 321 (6071):674–679.

67. Hood, L. 2008. A personal journey of discovery: Developing technology and changing biology. *Annual Review of Analytical Chemistry* 1:1–43.

68. Davies, K. 2001. *Cracking the Human Genome*. Free Press.

69. National Science Foundation: Science and Technology Centers: Integrative Partnerships. Available from: http://www.nsf.gov/od/oia/programs/stc/.

70. Hood, L. 2002. My life and adventures integrating biology and technology. Lecture for the 2002 Kyoto Prize in Advanced Technologies. Available at: http://www.systemsbiology.org/download/2002Kyoto.pdf.

71. Cook-Deegan, R. 1994. *GeneWars: Science, Politics and the Human Genome*. W.W. Norton & Company.

72. Office of Health and Environmental Research. 1986. Sequencing the Human Genome Summary Report of the Santa Fe Workshop. Department of Energy.

73. Major Events in the U.S. Human Genome Project and Related Projects. Available from: http://www.ornl.gov/sci/techresources/Human_Genome/project/timeline.shtml.

74. Lee, W. D. 2009. Interview with Eric Lander.

75. Mullis, K. B. 1990. The unusual origin of the polymerase chain reaction. *Scientific American* 262 (4):56–61, 64–65.

76. Fore, J., I. R. Wiechers, and R. Cook-Deegan. 2006. The effects of business practices, licensing and intellectual property on the polymerase chain reaction: Case study. *Journal of Biomedical Discovery and Collaboration*.1:7.

77. Henson, P. 1994. History of PCR. Smithsonian Video History Collection.

78. Foster, M. R. 1998. Guide to Thomas Kuhn's *The Structure of Scientific Revolutions*. Available at: http://philosophy.wisc.edu/forster/220/kuhn.htm.

79. Michl, H. 1951. Paper electrophoresis at potential difference of 50 volt/cm. *Monatshefte für Chemie* 82:489–493.

80. Muller, A. 1923. The X-ray investigation of fatty acids. *Proceedings of the Physical Society of London* 123:2043.

81. Perutz, M. F. 1963. X-ray analysis of hemoglobin. *Science* 140:863–869.

82. Davis, R. Stanford Genome Technology Center. Available from: http://med.stanford.edu/sgtc/.

83. Tener, G. M., P. T. Gilham, W. E. Razzell, A. F. Turner, and H. G. Khorana. 1959. Studies on the chemical synthesis and enzymatic degradation of desoxyribo-oligonucleotides. *Annals of the New York Academy of Sciences* 81:757–775.

84. Allewell, N. Report of the NIGMS Glue Grant Program Interim Outcomes Assessment Meeting. Available from: http://publications.nigms.nih.gov/reports/gluegrants/chairpersons_summary_of_expert_panel_recommendations.pdf.

85. Kaiser, J. 2011. Panel wants NIGMS to stop funding Glue Grants. *ScienceInsider* (May):27. Available at: http://news.sciencemag.org/scienceinsider.

86. Bio: Professor Robert Langer. Available from: http://web.mit.edu/langerlab/langer.html.

87. Herzenberg, L. A., D. Parks, B. Sahaf, O. Perez, M. Roederer, and L. A. Herzenberg. 2002. The history and future of the fluorescence activated cell sorter and flow cytometry: A view from Stanford. *Clinical Chemistry* 48 (10):1819–1827.

88. Herzenberg, L. A., and R. G. Sweet. 1976. Fluorescence-activated cell sorting. *Scientific American* 234 (3):108–117.

89. Acha-Orbea, H., R. M. Zinkernagel, and H. Hengartner. 1985. Cytotoxic T cell clone-specific monoclonal antibodies used to select clonotypic antigen-specific cytotoxic T cells. *European Journal of Immunology* 15 (1):31–36.

90. Crowley, M., K. Inaba, M. Witmer-Pack, and R. M. Steinman. 1989. The cell surface of mouse dendritic cells: FACS analyses of dendritic cells from different tissues including thymus. *Cellular Immunology* 118 (1):108–125.

91. Jones, P. P., S. W. Craig, J. J. Cebra, and L. A. Herzenberg. 1974. Restriction of gene expression in B lymphocytes and their progeny. II. Commitment to immunoglobulin heavy chain isotype. *Journal of Experimental Medicine* 140 (2):452–469.

92. Lee, W.D. 2009. Interview with Leighton Read.

93. Lenoir, T., and E. Giannella. 2006. The emergence and diffusion of DNA microarray technology. *Journal of Biomedical Discovery and Collaboration* 1:11.

94. About Affymetrix. Available from: http://www.affymetrix.com/estore/about_affymetrix/index.affx;jsessionid=B7ABF9F5E0AF12F894D47624FAF90033?category=34022&categoryIdClicked=34022&rootCategoryId=34003&navMode=34022&parent=34022&aId=aboutNav.

95. NIH. Report funding facts. Available from: http://report.nih.gov/fundingfacts/index.cfm.

96. Weinberg, A. M. 1961. Impact of large-scale science on the United States: Big science is here to stay, but we have yet to make the hard financial and educational choices it imposes. *Science* 134 (3473):161–164.

97. Sabin, A. B. 1967. Collaboration for accelerating progress in medical research. *Science* 156 (3782):1568–1571.

98. Esparza, J., and T. Yamada. 2007. The discovery value of "Big Science." *Journal of Experimental Medicine* 204 (4):701–704.

99. Lewin, R. 1986. Proposal to sequence the human genome stirs debate. *Science* 232 (4758):1598–1600.

100. Report on the Human Genome Initiative for the Office of Health and Environmental Research. Available from: http://www.ornl.gov/sci/techresources/Human_Genome/project/herac2.shtml.

101. House, W. The Human Genome Project. 2000.

102. Cook-Deegan, R. 1997. Does NIH need a DARPA? *Issues in Science and Technology* 13:25–28.

103. Gelijns, A. 2013. Manuscript review.

104. Gedeon, A. 2006. *Science and Technology in Medicine*. Springer.

105. Griffith, L. G., and A. J. Grodzinsky. 2001. Advances in biomedical engineering. *Journal of the American Medical Association* 285 (5):556–561.

106. First commercial technical information bulletin on NMR. Available from: http://www.scribd.com/doc/49350573/NMR-history-Varian.

107. Marton, L. 1979. *Advances in Electronics and Electron Physics*. New York: Academic Press.

108. Woo, D. J. 2006. A short history of the development of ultrasound in obstetrics and gynecology: Karl Theo Dussik Bio. Available at: http://www.ob-ultrasound.net/dussikbio.html.

109. Gelijns, A. C. 1995. *From the Scalpel to the Scope: Endoscopic Innovations in Gastroenterology, Gynecology and Surgery*. Sources of Medical Technology: Universities and Industry. Medical Innovation at the Crossroads, Vol. V. National Academies Press.

110. Rodbell, M. 1964. Metabolism of isolated fat cells. I. Effects of hormones on glucose metabolism and lipolysis. *Journal of Biological Chemistry* 239:375–380.

111. Northup, J. K., P. C. Sternweis, M. D. Smigel, L. S. Schleifer, E. M. Ross, and A. G. Gilman. 1980. Purification of the regulatory component of adenylate cyclase. *Proceedings of the National Academy of Sciences of the United States of America* 77 (11):6516–6520.

112. Sternweis, P. C., J. K. Northup, M. D. Smigel, and A. G. Gilman. 1981. The regulatory component of adenylate cyclase. Purification and properties. *Journal of Biological Chemistry* 256 (22):11517–11526.

113. Burnett, G., and E. P. Kennedy. 1954. The enzymatic phosphorylation of proteins. *Journal of Biological Chemistry* 211 (2):969–980.

114. Southern, E. 1975. Detection of specific sequences among DNA fragments separated by gel electrophoresis. *Journal of Molecular Biology* 98:503–517.

115. Stephens, P. J., P. S. Tarpey, H. Davies, P. Van Loo, C. Greenman, D. C. Wedge, et al. 2012. The landscape of cancer genes and mutational processes in breast cancer. *Nature* 486 (7403):400–404.

116. Kreeger, P. K., and D. A. Lauffenburger. 2010. Cancer systems biology: A network modeling perspective. *Carcinogenesis* 31 (1):2–8.

117. Vander Heiden, M. G., L. C. Cantley, and C. B. Thompson. 2009. Understanding the Warburg effect: The metabolic requirements of cell proliferation. *Science* 324 (5930):1029–1033.

118. Dell, H. 2006. Milestones 1: Observations from a ploughman. *Nature Reviews Cancer* 6:S7.

119. Hoption Cann, S. A., J. P. van Netten, and C. van Netten. 2003. Dr William Coley and tumour regression: A place in history or in the future. *Postgraduate Medical Journal* 79 (938):672–680.

120. Lathrop, A. E., and L. Loeb. 1916. Further investigations on the origin of tumors in mice. III. On the part played by internal secretion in the spontaneous development of tumors. *Journal of Cancer Research* 1 (1):1–19.

121. Nowell, P. C., and D. A. Hungerford. 1960. Chromosome studies on normal and leukemic human leukocytes. *Journal of the National Cancer Institute* 25:85–109.

122. University of Pennsylvania Health System. Legacy of the Philadelphia chromosome: Connecting chromosomes to cancer: Historic discovery. Available from: http://www.uphs.upenn.edu/news/features/philadelphia-chromosome/history/photo.html.

123. Martin, G. S. 1970. Rous sarcoma virus: A function required for the maintenance of the transformed state. *Nature* 227 (5262):1021–1023.

124. Stehelin, D., H. E. Varmus, J. M. Bishop, and P. K. Vogt. 1976. DNA related to the transforming gene(s) of avian sarcoma viruses is present in normal avian DNA. *Nature* 260 (5547):170–173.

125. Varmus, H. E., and P. R. Shank. 1976. Unintegrated viral DNA is synthesized in the cytoplasm of avian sarcoma virus-transformed duck cells by viral DNA polymerase. *Journal of Virology* 18 (2):567–573.

126. Sheiness, D., L. Fanshier, and J. M. Bishop. 1978. Identification of nucleotide sequences which may encode the oncogenic capacity of avian retrovirus MC29. *Journal of Virology* 28 (2):600–610.

127. Naumov, G. N., L. A. Akslen, and J. Folkman. 2006. Role of angiogenesis in human tumor dormancy: Animal models of the angiogenic switch. *Cell Cycle* 5 (16):1779–1787.

128. Nielsen, C. H., R. H. Kimura, N. Withofs, P. T. Tran, Z. Miao, J. R. Cochran, et al. 2010. PET imaging of tumor neovascularization in a transgenic mouse model with a novel 64Cu-DOTA-knottin peptide. *Cancer Research* 70 (22):9022–9030.

129. Brindle, K. M., S. E. Bohndiek, F. A. Gallagher, and M. I. Kettunen. 2011. Tumor imaging using hyperpolarized 13C magnetic resonance spectroscopy. *Magnetic Resonance in Medicine* 66:505–519.

130. Lutz, A. M., J. K. Willmann, F. V. Cochran, P. Ray, and S. S. Gambhir. 2008. Cancer screening: A mathematical model relating secreted blood biomarker levels to tumor sizes. *PLoS Medicine* 5 (8):e170.

131. Butler, T. P., and P. M. Gullino. 1975. Quantitation of cell shedding into efferent blood of mammary adenocarcinoma. *Cancer Research* 35 (3):512–516.

132. Sequist, L. V., S. Nagrath, M. Toner, D. A. Haber, and T. J. Lynch. 2009. The CTC-chip: An exciting new tool to detect circulating tumor cells in lung cancer patients. *Journal of Thoracic Oncology* 4 (3):281–283.

133. Yu, M., S. Stott, M. Toner, S. Maheswaran, and D. A. Haber. 2011. Circulating tumor cells: Approaches to isolation and characterization. *Journal of Cell Biology* 192 (3):373–382.

134. Klein, C. A. 2009. Parallel progression of primary tumours and metastases. *Nature Reviews Cancer* 9 (4):302–312.

135. Maheswaran, S., L. V. Sequist, S. Nagrath, L. Ulkus, B. Brannigan, C. V. Collura, et al. 2008. Detection of mutations in EGFR in circulating lung-cancer cells. *New England Journal of Medicine* 359 (4):366–377.

136. Liu, J. T., M. J. Mandella, N. O. Loewke, H. Haeberle, H. Ra., W. Piyawattanametha, et al. 2010. Micromirror-scanned dual-axis confocal microscope utilizing a gradient-index relay lens for image guidance during brain surgery. *Journal of Biomedical Optics* 15 (2): 026029.

137. Weinberg, R. 2007. *The Biology of Cancer*. Garland Science.

138. Waldmann, T. A., and J. C. Morris. 2006. Development of antibodies and chimeric molecules for cancer immunotherapy. *Advances in Immunology* 90:83–131.

139. Ferrari, M. 2010. Vectoring siRNA therapeutics into the clinic. *Nature Reviews. Clinical Oncology* 7 (9):485–486.

140. Senzer, N., and J. Nemunaitis. 2009. A review of contusugene ladenovec (Advexin) p53 therapy. *Current Opinion in Molecular Therapeutics* 11 (1):54–61.

141. Gabrilovich, D. I. 2006. INGN 201 (Advexin): Adenoviral p53 gene therapy for cancer. *Expert Opinion on Biological Therapy* 6 (8):823–832.

142. Goode, W. 1997. *Diagnosis and Treatment of Acute Promyelocytic Leukemia*. Memorial Sloan-Kettering Cancer Center.

143. Zhou, G. B., W. L. Zhao, Z. Y. Wang, S. J. Chen, and Z. Chen. 2005. Retinoic acid and arsenic for treating acute promyelocytic leukemia. *PLoS Medicine* 2 (1):e12.

144. Wang, Z. Y. 2003. Ham-Wasserman lecture: Treatment of acute leukemia by inducing differentiation and apoptosis. *Hematology* 1–13.

145. Rapp, U. R., M. D. Goldsborough, G. E. Mark, T. I. Bonner, J. Groffen, F. H. Reynolds, Jr., and J. R. Stephenson. 1983. Structure and biological activity of v-raf, a unique oncogene transduced by a retrovirus. *Proceedings of the National Academy of Sciences of the United States of America* 80 (14):4218–4222.

146. Hunter, T., and B. M. Sefton. 1980. Transforming gene product of Rous sarcoma virus phosphorylates tyrosine. *Proceedings of the National Academy of Sciences of the United States of America* 77 (3):1311–1315.

147. Sefton, B. M., K. Beemon, and T. Hunter. 1978. Comparison of the expression of the src gene of Rous sarcoma virus in vitro and in vivo. *Journal of Virology* 28 (3):957–971.

148. Hunter, T. History of tyrosine kinases. 1994; Available from: http://www.nih.gov/catalyst/back/94.07/Seminar.html.

149. Morrison, D. K., G. Heidecker, U. R. Rapp, and T. D. Copeland. 1993. Identification of the major phosphorylation sites of the Raf-1 kinase. *Journal of Biological Chemistry* 268 (23): 17309–17316.

150. Davies, H., G. R. Bignell, C. Cox, P. Stephens, S. Edkins, S. Clegg, et al. 2002. Mutations of the BRAF gene in human cancer. *Nature* 417 (6892):949–954.

151. Wan, P. T., M. J. Garnett, S. M. Roe, S. Lee, D. Niculescu-Duvaz, V. M. Good, et al. 2004. Mechanism of activation of the RAF-ERK signaling pathway by oncogenic mutations of B-RAF. *Cell* 116 (6):855–867.

152. King, A. J., D. R. Patrick, R. S. Batorsky, M. L. Ho, H. T. Do, S. Y. Zhang, et al. 2006. Demonstration of a genetic therapeutic index for tumors expressing oncogenic BRAF by the kinase inhibitor SB-590885. *Cancer Research* 66 (23):11100–11105.

153. Tsai, J., J. T. Lee, W. Wang, J. Zhang, H. Cho, S. Mamo, et al. 2008. Discovery of a selective inhibitor of oncogenic B-Raf kinase with potent antimelanoma activity. *Proceedings of the National Academy of Sciences of the United States of America* 105 (8):3041–3046.

154. Flaherty, K. T., I. Puzanov, K. B. Kim, A. Ribas, G. A. McArthur, J. A. Sosman, et al. 2010. Inhibition of mutated, activated BRAF in metastatic melanoma. *New England Journal of Medicine* 363 (9):809–819.

155. Nazarian, R., H. Shi, Q. Wang, X. Kong, R. C. Koya, H. Lee, et al. 2010. Melanomas acquire resistance to B-RAF(V600E) inhibition by RTK or N-RAS upregulation. *Nature* 468 (7326): 973–977.

156. Beck, T. W., M. Huleihel, M. Gunnell, T. I. Bonner, and U. R. Rapp. 1987. The complete coding sequence of the human A-raf-1 oncogene and transforming activity of a human A-raf carrying retrovirus. *Nucleic Acids Research* 15 (2):595–609.

157. Sharp, P. A., et al. The convergence of the life sciences, physical sciences and engineering. 2011; Available from: http://dc.mit.edu/sites/dc.mit.edu/files/MIT%20White%20Paper%20on%20Convergence.pdf.

158. Pennisi, E. 2003. A hothouse of molecular biology. *Science* 300 (5617):278–282.

159. Varmus, H. Provocative questions. Available from: http://provocativequestions.nci.nih.gov/.

Index